Talk Human to Me

Dr. Christopher S. Vogelmann

Talk Human to Me

Make AI Your Wingman, Not Your Replacement

Kindle ISBN: 978-1-967459-03-2
Paperback ISBN: 978-1-967459-04-9
Hardcover ISBN: 978-1-967459-05-6
Audiobook ISBN: 978-1-967459-06-3

Important Notice

Disclaimer:

This book is intended to provide educational information on the subjects discussed. The content provided is for informational purposes only and should not be construed as legal, accounting, or professional advice. The author and publisher are not engaged in providing legal, accounting, or any other professional services. If legal advice or other expert assistance is required, the services of a qualified professional should be sought.

The strategies, tips, and tools discussed in this book are based on the author's personal experience and knowledge in the field of AI and digital marketing. The outcomes discussed are not guaranteed and may vary depending on individual circumstances. The author and publisher expressly disclaim any warranties, whether express or implied, including, but not limited to, warranties of merchantability, fitness for a particular purpose, and non-infringement. The author and publisher shall not be liable for any losses or damages, including, but not limited to, direct, indirect, incidental, special, or consequential damages arising out of or in connection with the use of this book.

Please note that the digital marketing and AI landscape are constantly evolving. While the strategies and insights shared in this book are based on the latest trends and practices at the time of writing, it is essential to stay informed and adapt to new developments to

Printed in the United States of America

For the humans who get curious before they get cautious,

take action before they feel ready,

and never let the robots have all the fun.

Contents

Talk Human to Me

Make AI Your Wingman, Not Your Replacement

It started like a scene from a sitcom: my friend was chatting back and forth with his AI tool like it was his sassy, overachieving roommate. Meanwhile, I sat there—half fascinated, half in an existential tech crisis.

He was breezing through a dozen tasks with the kind of ease that suggested he'd either made a deal with the machine, or he genuinely enjoyed reading instruction manuals. For fun.

Me? I was the one still arguing with my printer. (Spoiler alert: the printer always wins.)

And in that moment, the thought hit me: *"This is how it ends. Death by irrelevance."*

For a split second, I wondered if I'd officially hit the point where the world started speaking a language I didn't.

I wish I could say that moment was unusual. It's not.

The future isn't coming. It's here and already copying your homework.

AI has leapt out of sci-fi and straight onto our kitchen counters, commuter playlists, and kids' homework. One minute we're watching robots in movies, the next your phone is writing performance reviews better than most humans can.

Suddenly, everyone's supposed to have an "AI strategy" based on algorithms most people can't pronounce, let alone trust. It's like being handed the controls to a space shuttle and told, "Cool, now land it before lunch."

Why I Wrote This (and Why You Might Need It Too)

I wrote *Talk Human to Me* because I've seen what happens when smart, capable people start to second-guess themselves.

I've watched the panic when someone opens an AI app and freezes. The polite smiles during jargon-soaked webinars. The hopeful eyes when someone promises, "Don't worry. No coding required." (Translation: Prepare for mild to moderate suffering.)

Here's the truth: The pace of change has outpaced most people's comfort zones. And somewhere along the way, we got sold a lie—that if you're not already a tech expert, you're too late to the party.

That's garbage.

You already have everything you need. It's just buried under a pile of overwhelm, outdated beliefs, and way too much conflicting advice from people who've never struggled with technology.

If your first instinct is to find AI's off switch and hide it behind the blender—this book is for you.

What Makes This Different

This isn't a tech manual. You won't find code, quizzes, or condescending techspeak here.

You will find small wins, clear steps, and maybe even a little joy along the way.

Because here's the secret no one's telling you: AI should work for you, not the other way around.

This isn't about becoming a digital wizard. It's about becoming more you—with tools that finally speak your language. Real human language. With commas and questions and moments where you pause and say, "Huh. That's actually kind of cool."

Inside, you'll discover:

- **Why AI feels so scary** (and why that reaction's completely normal)
- **What AI really is**—and what it definitely isn't
- **How to shift your mindset** without losing your mind
- **Your first five tools** that actually work (no coding, no crying)
- **Simple strategies** to stay grounded while the world speeds up

The transformation you'll experience isn't just about tools. It's about power. Not the loud, flashy kind. The quiet kind. The kind that whispers, *"You've got this."*

Who Am I? (And Why Am I Still Talking to Printers?)

I didn't start in a lab or a think tank. I started at my kitchen table, cursing at that damn blinking cursor and wondering why every tutorial assumed I could decode tech before breakfast.

At first, it was messy. Post-Its covering the walls (basically wallpaper). Late-night scribbles. Trial and lots of error. But every time something clicked, I turned it into a guide.

Eventually, the skeptic in me became the strategist.

I studied psychology and social relations at Harvard, where I had monthly dinners with Professor B.F. Skinner and worked as a research assistant to Professor David McClelland, the father of need theory and achievement motivation.

Skinner taught me what works in the messiness of real life. He didn't care about theory for theory's sake; he cared about behavior that actually changed things. He was highly ethical, surprisingly emotional, and deeply human.

McClelland taught me why humans need what they need. His work on achievement motivation revealed that understanding *what drives people* is just as important as understanding what they do. He was warm, empathetic, and genuinely curious about what made individuals thrive.

One mentor focused on behavior. The other on motivation. Together, they gave me the lens I use today: **practical tools delivered with genuine empathy.**

That's how I approach AI: through what actually works for real humans in real situations, while honoring what makes us human in the first place.

For the past five years, I've worked with business owners, creators, and everyday humans to turn overwhelm into confidence. Not with hype, but with real tools and clear thinking.

My secret weapon? I never forgot what it felt like to be on the outside looking in. That makes me a better guide, and a much more patient teacher.

Here's Your Invitation

This book isn't about perfection. It's about permission.

Permission to fumble and laugh at the learning curve. To question the hype and demand something that actually works for you.

You don't need to be first, fastest, or fearless. You just need to take one human-sized step.

If you're ready for that, we're already in this together.

Chapter 1
Why AI Feels So Scary
(and Why You're Not Crazy)

> *"You're not broken. The world just updated without your permission."*

The first time I felt that creeping "oh-no-what-now" dread about AI, like the universe had silently pushed a software update on the entire human experience while I was still trying to figure out how to cancel a Zoom trial, I thought of my mom.

This woman could bake seven-layer German cakes from memory, speak seven languages, and negotiate with telemarketers like a trained hostage negotiator.

But put her in front of a McDonald's touchscreen kiosk? She froze like she was being asked to defuse a bomb with a pencil. The confused squint. The nervous chuckle. The slow backpedal away from the glowing rectangle as if it might suddenly demand her password and firstborn child (me).

And now I see it everywhere: that same expression of tech-induced vertigo on smart, capable, successful grown-ups. You know, the kind who can run a business, file their taxes, and wrangle toddlers, but who crumble when asked to "log in with Google."

Let's be clear: I didn't exactly waltz into the AI revolution with unbridled swagger myself.

The first time I opened ChatGPT, I spent five minutes overanalyzing how to say hello.

Was "Hi" too casual? Was "Hello" too formal? Would it judge me if I used an emoji? (And yes, I'm a certified emojiholic.)

Eventually, I remembered I wasn't talking to HAL 9000. I was talking to a glorified autocomplete. And honestly? That was the first time I exhaled and thought, "Okay. I can do this."

Here's the Real Deal

It's not that this stuff is new. It's that it's all new, all at once, all the time. Like life suddenly hit 2× speed, without checking if your brain had enough RAM.

You go to sleep a competent human, and by morning your job title's changed, your software's out of date, and someone on LinkedIn is casually posting about how she built a business empire overnight using three prompts and a Canva template.

If your nervous system feels like it's been run over by a digital bulldozer, congrats. That means you're human. And your brain is working exactly as it should.

See, your brain wasn't built for algorithmic evolution. It was built for planting season. You know, sunrise, sunset, maybe the occasional sabertooth tiger. Not "learn Midjourney by Tuesday or risk irrelevance."

That anxiety? It's not a personal failure. It's not a tech allergy. It's your brain's fire alarm going, "Uhh, is our species still safe here?!"

You've Done This Before

We've been here before. The printing press terrified the monks. Radio made people nervous. Television was going to rot our brains (okay, maybe that one wasn't entirely wrong).

But here's what those skeptics all had in common: they adapted. And you did too.

You've already gone from rotary phones to FaceTime. From fax machines to Slack. Used to print MapQuest directions. Now the car tells you where to turn and throws shade when you miss the exit.

This isn't your first tech rodeo. It's just your fastest one.

And that's where the fear creeps in. Because speed feels like chaos when no one handed you a map.

So let's make one.

If You've Ever...

If you've ever:

- Broken into a cold sweat when someone says "prompt engineering"
- Felt deep shame because a 24-year-old asked, "Wait, you're not on Midjourney?"
- Read a post about AI replacing consultants and briefly considered living off the grid in a yurt

You're my people.

This book isn't here to calm you down with platitudes. It's here to plug you back into your power.

What You Actually Need

Here's the truth no one's yelling over the AI hype parade: You don't need to master machine learning. You don't need to memorize every tool.

You just need to stay a little more curious than afraid.

Your curiosity is your survival instinct in a sequined blazer. It's what keeps you asking, "Wait, what if I could learn this?"

You've done impossible things before. You've survived awkward tech rollouts, online learning, onboarding videos with bad smooth jazz soundtracks. You've figured out taxes. Relationships. Roombas.

You. Can. Learn. This.

That flutter in your gut? That's not panic. **That's potential doing a soundcheck.**

It's About Identity, Not Information

This chapter isn't about information. It's about identity.

You're not obsolete. You're not late. You're just... updating.

I had dinner once a month with B.F. Skinner, the famous behaviorist, and even he would be fascinated (and mildly horrified) by how fast reinforcement loops now happen. He used pigeons. We use push notifications.

But the principle remains: behavior adapts when the right conditions exist. You've adapted before. You'll adapt again.

So let's do this together.

Welcome to the chaos. Let's start making it work for you.

Chapter 2
Riding the Tsunami of Change

"Fear is just curiosity dressed in yesterday's clothes."

You know that moment when something big sneaks up behind you—unannounced, slightly damp, maybe holding a spreadsheet—and your instincts fire like it's go time?

Fight. Flight. Or fling a stapler and pretend it's your stress-ball substitute.

Yeah, that's your ancient brain doing exactly what it was built for: saving your gorgeous, well-meaning self from lions, cliffs, and unexpected tax audits.

Now the lion's a chatbot, and the cliff's labeled "technological disruption."

Let's be honest: your stomach does a little backflip when you hear that phrase, right? Totally normal. That's not weakness. **That's your amygdala booting up ancient code to deal with a world it never imagined would include Deepfake Drake and AI-powered PowerPoints.**

See, your brain was built for sunrises, seasonal harvests, and the occasional spear-wielding neighbor. Not this never-ending avalanche of digital transformation.

No one emerged from the Stone Age holding an iPhone. But here you are: texting your dentist, Zooming from your kitchen, Googling "how to feel less like a dinosaur," and managing to remember at least two streaming passwords.

You've Done This Before

We've been through revolutions before. Electricity. Radio. The fax machine (RIP).

And every single time, the world panicked, adapted, and then pretended it was totally fine the whole time.

Adaptation isn't new—just never this caffeinated.

But let's take a breath. Inhale. Exhale. Laugh at yourself if it helps. Because you've already adapted more times than you give yourself credit for.

Think about it: The friend who once swore she'd never use GPS now asks Siri where the nearest oat-milk latte is; your aunt who once called texting "impersonal" now sends you blurry cat memes at 7 a.m.; your skeptical coworker is now the office's unofficial Slack-emoji sommelier.

The adaptation cycle looks like this:

Skepticism → Resistance → Reluctant trial → Casual dependence → Unshakable loyalty

That's the cycle. It's human. It's messy. It works.

When Panic Creeps In

But somewhere between "What is this AI thing?" and "Oh no, it's coming for my job," the panic creeps in like a nosy neighbor with binoculars and opinions.

You hear whispers in your own mind:

- "I'm too old for this"
- "It's too technical"
- "Maybe it's just a trend?"
- "If I learn it, am I helping the robot that replaces me?"
- "I'll just wait this one out. Maybe it'll go away"

Real talk: It won't. **AI isn't a fidget spinner. It's more like electricity, with a to-do list and a LinkedIn profile.**

Common Mental Traps

Trap #1: Believing it's not your field.

As if innovation only knocks on the doors of people wearing hoodies and pitching apps to VCs.

Trap #2: Telling yourself you're too old.

This, from the same folks juggling three remotes while FaceTiming their grandkids and finding memes faster than a teenager with 5G.

Trap #3: Waiting for perfection.

The first time you drove, it wasn't perfect. The first time you made coffee, it tasted like regret. But you kept going.

Truth bomb: Adaptation is not a youth thing. It's a curiosity thing.

Curiosity doesn't age. It just updates.

That spark you think you've lost? It's just waiting for a reason to stretch.

Age Counts Years. Curiosity Counts Miles.

The people who ran the world before the internet now correct their grandkids' grammar in group texts.

The most adaptable folks? They poke at things. They click buttons they don't understand (with mild terror). They try weird apps, ask dumb questions, laugh when they mess up, and do it again.

They don't master it first. They muddle through with flair.

The Mastery Myth

Let's squash this toxic belief like a bug in a broken Roomba: *"I need to understand how AI works before I use it."*

Oh really? Do you understand the full mechanics of a microwave but still use it to reheat your pizza? Thought so.

This isn't about becoming a Prompt Jedi. It's about building momentum, not waiting for perfection.

While you hesitate, others are experimenting. While you overthink, others are prompting, playing, publishing, and profiting.

This isn't about being first. It's about not being frozen.

Choosing not to engage isn't a pause. It's a silent surrender.

Your Experience Still Matters

Your value doesn't vanish because the tools evolve. In fact, all the reasons you think you can't adapt? That's why you're essential.

Your experience. Your discernment. Your healthy suspicion of hype.

AI doesn't have that. It has data. You have discernment. It guesses. You know.

So next time you get that itchy feeling in your gut, like maybe it's time to run or fake your own retirement party, remember: **Fear means your attention is exactly where it needs to be.**

You don't need to outrun the wave—just learn to move with it.

The tide's rising. You've got balance, history, and one hell of a surfboard.

Time to ride.

Chapter 3

The Four Natural Responses to Tech Fear

"AI doesn't take prisoners. It takes patterns. Change yours."

Picture this: the first time a blinking VCR clock demanded to be set, I wanted to throw the remote across the living room and possibly set the whole device on fire.

That pulse of murderous rage wasn't about the technology. It was my unregulated nervous system throwing a full-blown tantrum because some plastic rectangular black box was making me feel like a medieval peasant confronting a spaceship.

Every time a new wave of innovation crashes into our carefully constructed routines, some prehistoric part of our brain lights up like a casino alarm, ready to fight a saber-toothed tiger, flee to the nearest cave, freeze like roadkill, or—on spectacularly rare days—surf the chaos like we actually know what we're doing.

I used to think my love/hate relationship with anything newer than a rotary phone made me a technological Neanderthal. Turns out, I'm just like you: a wonderfully predictable human.

The Four Tech Threat Reflexes

We all have a default response when new technology threatens our comfort zone. Each one comes with its own performance. Once you know your default drama, you stop blaming "stupidity" or "laziness" and start seeing the pattern for what it is: perfectly normal, occasionally hilarious, and surprisingly hackable.

Fight Mode: The Inner Tech Gladiator

If you've ever declared, "AI is garbage and I want nothing to do with it!" while secretly googling "AI for beginners" at 2 a.m. in incognito mode, you've met your inner Tech Gladiator.

This response is all righteous fury with zero curiosity. Fight feels deliciously superior. It's excellent at preserving the status quo, even when that "status quo" expired five software generations ago.

The Gladiator's motto: "Back in my day, we didn't need seventeen passwords to check our email!"

Flight Mode: The Digital Escape Artist

"I'll deal with this... someday," you announce with the confidence of someone who genuinely believes "someday" exists in any meaningful timeline. Then you slide your phone under the pillow so you don't have to see those increasingly passive-aggressive "Update Now" notifications.

Procrastination becomes high art. The only thing you update with any regularity is your excuse for not updating.

The Escape Artist's motto: "Maybe if I ignore it long enough, it'll just work itself out?"

Freeze Mode: The Paralysis Perfectionist

Welcome to the endless scroll of doom. The towering mountain of bookmarked "how-to" articles that you'll definitely read "when you have time."

You call it "research," but really, you've entered the digital equivalent of stage fright: standing in the spotlight with your mouth open, hoping someone will throw you a line.

I've personally lost hundreds of hours in this wasteland, convinced that one more review, one more tutorial, one more expert opinion would finally tell me which button to click without triggering the apocalypse.

Five thousand browser tabs later, it's still just fear wearing a very convincing "I'm being thorough" costume.

Flow Mode: The Unicorn State

This one's rarer than a polite comment section, but oh so good. Here, there's no panic, no drama, no existential crisis. Just curiosity in comfy clothes.

You poke things. You stumble spectacularly. You laugh at your mistakes instead of having a small breakdown. You learn stuff.

Flow isn't fear-proof. It's just fear-tolerant, like that friend who can eat spicy Buffalo wings without crying.

Spot Your Default

Most of us have a primary reflex when something shiny and new threatens our carefully curated ignorance. Some of us are over-achievers and rotate through all four modes depending on our caffeine levels and whether Mercury is in retrograde.

The trick is catching yourself mid-performance.

For me, Fight shows up when tech threatens my delicate sense of competence. Flight kicks in when the "new thing" feels like pointless digital busywork. Freeze commandeers the wheel when options multiply like rabbits and my brain goes into full meltdown mode.

For you, the cast of characters might be different. The point isn't to audition for the "best" response. It's to recognize your signature move. **That's when the shame starts its slow, beautiful retreat.**

You're not bad at tech. Your brain is just running Stone Age software in an iPhone world.

Here's your assignment: start a Tech Fear Diary. Every time you dodge an app, rage at a new interface, or scroll yourself into a Wikipedia coma, jot down which flavor of panic you're experiencing.

Is your inner Gladiator swinging wildly? Your Escape Artist faking their own death? Your Perfectionist playing dead until the scary technology goes away?

Awareness means the reflex can't run the show indefinitely. You become the director instead of just another actor stumbling through the same tired script.

You've Done This Dance Before

You've stared tech fear directly in its pixelated face and lived to tell the tale, probably more times than your selective memory wants to admit.

The first time you survived a video call without accidentally showing everyone your unmade bed. At some point, what started as "absolutely not, over my dead body" became "I can handle this before my second cup of coffee."

Make a list. Write down those moments when you went from Digital Disaster to Competent Human. Maybe you conquered online banking when your checkbook mysteriously vanished.

Each victory is a receipt. Each is hard evidence that your brain already knows how to shift from Fight, Flight, or Freeze to Flow.

Small Bets: The Raccoon Method

To escape the gravitational pull of tech dread and inch toward Flow, try what I call "tiny experiments." No commitments. No public humiliation. No risk of accidentally breaking the internet.

Generate some gloriously ridiculous AI art. Chat with a bot about your deepest philosophical questions. Follow exactly one step of a tutorial, then quit with zero guilt.

Think of it as clicking buttons like a curious raccoon with Wi-Fi. The goal isn't digital enlightenment. It's to prove to your nervous system that the world won't actually end if you mess up.

If Fight flares up ("This is a waste of my time!"), just note it like a mildly interesting weather report. If Flight suggests "Maybe later," give yourself exactly five minutes. If Freeze kicks in, switch to something physical and circle back when your brain reboots.

These little reps build comfort, not expertise. And comfort is more than enough.

You're Not Broken, You're Evolving

Here's what people don't want to admit: thinking you're broken because you struggle with technology is like thinking you're broken because you can't speak fluent Mandarin after hearing it once on a cooking show.

Stress and resistance are the default human responses during transitions, especially when change messes with your sense of control, competence, or fundamental worth.

So if you feel like a technological fraud, congratulations. You're following a script humans have been performing since the first person looked at fire and thought, "Well, this seems unnecessarily complicated."

Instead of layering shame on top of anxiety like some sort of emotional lasagna, treat the reaction for what it actually is: your brain doing its job, just using embarrassingly outdated software.

Once you see the pattern clearly, it becomes less tragic personal failing and more comedy gold.

Making Adaptation Less Accidental

You don't have to achieve Flow Mode enlightenment and maintain it 24/7. That's a fantasy reserved for Buddhist monks and people who actually read user manuals for fun.

The real goal is simpler: pause when the panic hits, name your current starring role, and make one tiny choice that breaks your pattern.

These micro-shifts are like scaffolding for your sanity. Over time, they become your new default. Less drama, more curiosity, significantly fewer angry monologues directed at inanimate objects.

Adaptation isn't about developing superhuman technical skills. It's about making enough micro-shifts to bore your fear into submission through sheer repetition.

Tech anxiety isn't a bug in your system. It's a feature, a delightfully stubborn human one.

Once you name your patterns, you realize something revolutionary: you don't actually have to obey them. You can observe, laugh, gently redirect.

Fight, Flight, and Freeze aren't villains. They're just opening acts. Flow is your permission slip to change the channel.

You've done this evolutionary dance before. You'll do it again, probably while complaining creatively the entire time.

Try a risk so microscopically small it's almost embarrassing. Notice how quickly shame packs its bags when curiosity shows up uninvited.

Pick your flavor. Then click. You're back in control.

Chapter 4
Mind the Gap
(And Then Close It)

There's a moment (sometimes whisper-quiet, sometimes screamingly obvious) when you realize the people around you have started doing more with suspiciously less effort.

They fire off replies faster than seems humanly possible. Ideas pour out like their imagination's on a Red Bull IV drip set to turbo. Their content, pitches, and systems hum along while you're still hunting for the perfect synonym in your mental thesaurus.

They're not smarter than you. (Trust me, I've met some of these people at conferences. They are most decidedly mortal.) They're not working longer hours. (Half of them are posting beach photos while somehow maintaining their LinkedIn thought leadership status.) They're just using AI like it's their unpaid but extremely competent intern.

That's the gap. And it's probably wider than you think.

How the Gap Opens

The gap opens slowly at first, like that one friend who suddenly became "really good at photography" after buying a phone with a decent camera. Then overnight, it becomes a chasm. Not between the tech-savvy wizards and the proudly analog holdouts, but between the quietly curious and the righteously resistant.

You don't need to become a technologist to close it. You just need to pay attention to who's gaining traction, and, more importantly, how they're doing it without breaking a sweat.

And suddenly you realize you're not falling behind on effort — just on experimentation.

The Telltale Signs You're Watching from the Sidelines

We tend to spot it too late, usually while muttering something about "kids these days":

The LinkedIn Overachiever: A colleague who suddenly writes 10 thoughtful posts before your first cup of coffee, each one somehow perfectly crafted and annoyingly insightful.

The Effortless Entrepreneur: A competitor who launches a polished offer while you're still having an existential crisis about whether your logo font conveys "trustworthy professional" or "trying too hard." (Personal experience.)

The Mysteriously Calm Creator: A friend who's suddenly become a "thought leader" and seems oddly relaxed about it, like they've discovered the secret to content creation without an accompanying mental breakdown.

It's not magic. It's leverage.

It's also mildly infuriating if you're still doing everything the hard way.

While you're overthinking the perfect hook for that email, they've got AI spitballing 20 drafts in the time it takes you to decide between "Hi" and "Hello."

While you reread that same paragraph for the 12th time (because maybe this time you'll catch the perfect phrasing), they've already shipped it, moved on, and somehow scheduled their next week of content during their lunch break.

While you're in a committed relationship with Canva, trying to make your graphics look less like they were designed by a caffeinated squirrel, they're already mapping full funnels.

And here's what I really want you to understand: the gap isn't actually a tech gap. **It's a willingness gap.** It's about your desire to adapt and change.

The Uncomfortable Truth About Early Adopters

The early adopters aren't always the smartest folks in the room. (I've been in those rooms. Some of them can barely operate a coffee machine without detailed instructions.) They're just the ones who shrugged and said, "Let's see what happens if I press this button."

Their secret? They play.

They didn't wait for permission from the universe, their boss, or their imposter syndrome. They gave themselves 30 minutes to play around. Then another 30. Then the beautiful, slightly unfair magic of compounding kicked in.

You don't need to catch up to where they are now. You just need to re-enter the game. Jump lanes and build your own momentum.

Step One: Detective Mode (Become a Student of Sudden Productivity)

Look around your industry, community, or even your inbox with fresh eyes. Who suddenly seems unusually effective, suspiciously prolific, or mysteriously calm about their workload?

That person who used to struggle with deadlines and now ships content like a machine with perfectly calibrated output settings? Probably not a productivity epiphany or a personality transplant. Probably an AI sidekick secretly doing the heavy lifting.

The overworked small agency that's now punching way above its weight class with killer branding and suspiciously consistent outreach? They probably offloaded the boring grunt work to something with a .ai domain and a cheerful willingness to work weekends.

Your Mission: Start paying attention to the results you secretly admire. Then trace the workflow backward. Not to copy their exact process (that's creepy and probably won't work for you anyway), but to decode the leverage they're using.

Step Two: Find Your Friction (Where Stress Meets Opportunity)

You don't need AI for everything. Just the parts that slow you down, wear you out, or trigger your perfectionism into its familiar death spiral.

Hate first drafts? Let AI be your brainstorming buddy who never judges your terrible initial ideas and actually makes them better. It's *so* good at that. And it's one of my favorite things to do.

Stuck writing bios, blurbs, or intros? That's literally a 10-minute prompt away from becoming a non-problem.

Dread email follow-ups like they're root canals? Automate the structure and save your human brain power for personalizing the heart of the message.

Feel like social media is a treadmill from hell designed by attention-seeking demons? Batch it with a voice-to-post workflow and reclaim your sanity.

The point isn't to replace your brain with a robot. It's to relieve your brain from the busywork, bottlenecks, and brain fog that keep you from doing your actual best work.

Your brain thrives on novelty. Curiosity literally rewires stress into focus.

Your attention is better spent where your experience, intuition, and irreplaceable human weirdness shine. Let AI carry the water so you can serve the wine.

Step Three: Study the Underdogs (They're Showing You the Way)

Big companies with massive teams and endless meetings will always move like glaciers. They move slowly and with a lot of unnecessary drama. It's the solopreneurs, the micro-agencies, the two-person bakeries with oddly professional Instagram Reels who reveal the real, practical use cases.

Case Study: The Neighborhood Bakery That Gets It

That little bakery down the street? They used ChatGPT to write their monthly newsletter (goodbye, staring at blank pages for three hours). Midjourney helped them design their seasonal cookie menu (goodbye, expensive graphic designer for simple layouts). ElevenLabs narrated their new "Behind the Dough" video series (goodbye, hiring a voice-over artist for their quirky little performances).

They're not chasing trends. They're baking more cookies and burning out less.

Revolutionary? No. Smart? Absolutely.

You don't need to become a coder who speaks in mysterious algorithms. You just need to become curious about what's possible.

Step Four: Play Without Pressure (Low Stakes, High Learning)

The gap doesn't close with panic-driven cramming sessions at 2 a.m. It closes with consistent, low-stakes practice.

The 20-Minute Experiment: Set a timer. Try something new with AI. Doesn't matter what. Seriously, you're not performing for anyone.

Write a product description for an imaginary gadget. Generate 10 different email subject lines. Ask for metaphors about your industry. Tell it to make you laugh about your biggest work frustration. Ask it to explain cryptocurrency like you're a golden retriever. (Try it. The answer will be hilarious.)

Whatever. Just practice the muscle of experimenting without needing a perfect outcome or a LinkedIn post about your "AI journey."

The people pulling ahead aren't using AI perfectly. They're not AI whisperers with secret prompting techniques. They're just using it. Imperfectly. Repeatedly. Creatively. With the same energy you'd bring to trying a new recipe: curious, not precious.

You're already wired for adaptation. It's what kept your ancestors alive and what will keep you relevant.

You can do this too. (Really. I've seen people who still call it "the Google" figure this out.)

Your Only Real Risk? Staying Comfortably Stuck

Or comfortably numb, if you prefer a Pink Floyd reference.

This is the moment when you can shift from feeling like you're falling behind in some imaginary race to actually mastering a game that's only just started.

You don't need a team of engineers, a computer science degree, or a trust fund to hire consultants. You need one tool, one workflow, one curiosity-fueled experiment that you're willing to try without over-thinking it to death.

Want leverage? Name one task you dread this week. Just one. (We're going to keep this stupidly simple.)

Now ask yourself: What if AI could help me do this faster, better, or, perhaps, not at all?

That's not wishful thinking or Silicon Valley hype. That's how quiet AI co-pilots become your unfair advantage, compounding progress while everyone else is still debating whether this whole "artificial intelligence thing" is just a passing fad.

Real talk: AI's not going anywhere. But neither is your ability to adapt, experiment, and maybe even enjoy the process.

The gap is real. But it's not permanent, and it's definitely not personal.

Pick something small. Try it badly. Try it again slightly less badly.

And hey. You don't need to stop using Canva. You just need to stop letting that overly-caffeinated squirrel choose your fonts.

Watch the gap close itself — one ridiculous, curiosity-fueled experiment at a time.

That's how revolutions start. Quietly, one click at a time.

Chapter 5
Your Tech Evolution Résumé
Proof You're Already Kicking Tech's Butt (and Forgot to Notice)

Let's be honest: you probably don't give yourself enough credit. You've evolved from rotary phones to voice texting your niece about the family group chat that mysteriously disappeared (again).

You've gone from printing out MapQuest directions to yelling at Google Maps when it "recalculates" mid-turn. You stream documentaries at double speed while buying dog food on Amazon and silently judging someone's email font choice.

But still, you say:

> "I'm not a tech person."

Really? Then who figured out how to mute themselves on Zoom faster than a teenager slamming their bedroom door?

Who reset their Wi-Fi router like a NASA technician under pressure?

Who FaceTimed their vet to show them their dog's mysterious rash, and still managed to hold the camera steady?

You.

You've done all that. And probably while sipping lukewarm coffee and Googling "why is my phone doing that thing again."

It's time to update your internal résumé.

What Is a Tech Evolution Résumé?

It's not for recruiters. It's not going on LinkedIn. It's for you.

It's the proof that you've already walked through the digital fire. And yes, sometimes you were wearing metaphorical flip-flops and cursing the whole way, but you made it through. This résumé exists to shut down that outdated mental software whispering,

> "You're behind."
>
> "You're too old."
>
> "You'll never catch up."

Check this: That voice is wrong. You've been adapting all along. You just never paused long enough to realize it.

Your Greatest Hits (That You Forgot Were Amazing)

Let's jog your memory.

Remember when you first:

- Switched from printing photos to uploading them (and panicked that you lost Grandma's birthday forever)?
- Paid a bill online instead of mailing a check (and then checked your bank account 12 times)?
- Used voice-to-text and accidentally sent "Meet me at five tacos" instead of "Meet me at 5 o'clock"?
- Bought your first airline ticket from your laptop without calling a travel agent? (OK, maybe you're not *that* old.)

You adapted. You learned. You didn't explode. That's résumé material.

So why do you still doubt your ability to handle AI?

Because you've never paused to total up the proof.

Build Yours: A Four-Question Framework

Here's how to map out your very own Tech Evolution Résumé. Pick any piece of tech you use now that once made you want to throw it across the room. Then answer these:

1. What totally freaked you out at first?

Example: "If I press the wrong button, will my bank account implode?"

2. How long did it take before it felt normal?

Be honest. "Three weeks and one YouTube video titled 'Google Drive for Dummies.'"

3. What helped you get over the hump?

A patient teenager? Rage Googling? Bribing yourself with snacks?

4. How does it feel now?

"Smooth." "Confident." "I can't believe I was scared of this."

Let's run a full example.

> **Tech:** Mobile Check Deposit
>
> **Initial freak-out:** "What if I deposit the same check twice and go to federal prison?"
>
> **Time to adapt:** "Two tries and a mild panic attack."
>
> **Bridge moment:** "Realizing the app tells you if it's a duplicate."

Now: "Easy. I deposit checks from my couch like a boss."

There. One entry down. You're already on a roll.

Why This Works

Your brain responds to evidence. When you see a list of times you survived (and even thrived) using technology, your fear response takes a back seat. Suddenly, AI isn't a mysterious threat. It's merely the next tool you'll probably mispronounce before you master it.

By the way… This chapter isn't about nostalgia. It's about reclaiming your narrative.

You've done this before. You're doing it now. And you'll keep doing it, whether the next leap is AI, augmented reality, or programming your smart fridge to stop buying oat milk.

Add AI to Your Résumé (Yes, Even Now)

Your next line item might look like this:

Tech: AI writing tools

Initial freak-out: "Will it take my job, or just write awkward poetry about it?"

First experiment: "Asked it to write a limerick about my business. It was… not great. But weirdly helpful."

Micro-win: "Used it to summarize a blog post. Saved 30 minutes. Felt like magic."

It doesn't have to be perfect. It just has to be yours.

You're building momentum. And the compounding effect of trying, failing, and trying again is exactly how tech confidence is born. Think play, not panic.

Final Thought: Confidence Doesn't Come from Tools

It comes from adding up your wins.

Nobody cares if you can build an app from scratch or jailbreak your iPhone with your mind. What matters is this:

- You've adapted when it wasn't easy.
- You've kept learning when it would've been easier to quit.
- You've survived (and often laughed through) the chaos.

And now you're here, building your Tech Evolution Résumé like the digital shape-shifter you didn't know you were.

You don't need a certificate. You only need a reminder.

So create a reminder for yourself. Print it. Frame it. Doodle on it. Post it near your desk as a permanent mic drop for your inner critic.

And the next time someone says,

> "Wow, you're good with tech!"

Don't flinch. Just smile and say,

> "I've got the résumé to prove it."

Chapter 6
Dismantling the AI Anxiety Myths

Time for a little Q&A, eh? (That was for my Canadian friends.)

Welcome to the part of the journey where we torch the lies, myths, and straight-up nonsense that keep smart people stuck in tech paralysis, whether you're running a business from your kitchen table or sitting in a cubicle wondering if your job will exist next year.

This is where we take a crowbar to your AI anxiety and pry it loose from the story it's been whispering in your ear: "You're too late. You're not smart enough. This is only for the Silicon Valley hoodie crew with venture capital and unlimited coffee budgets."

Wrong.

Let's get real, Dr. V-style, with eyebrow-raising honesty, a little bit of rebellion, and enough sass to short-circuit the self-doubt machine.

Funumental Rule #47: *Fear fades faster when you laugh while learning.*

MYTH #1: "I have to use every AI tool or I'll fall behind."

The Employee Version: "My coworker mentioned 12 different AI apps in one meeting. I'm doomed."

The Business Owner Version: "My competitor's website says they use AI for everything. I need to catch up immediately."

The Reality: Lie.

You don't need a PhD in prompt engineering or 57 browser tabs filled with Chrome extensions named after mythical creatures. You need one tool that takes something annoying off your plate. That's it.

One solved frustration beats ten half-baked experiments.

For Employees: Pick the task that makes you want to hide in the supply closet. Email drafts? Meeting summaries? Research rabbit holes? Start there. Master one thing before you add another.

For Business Owners: Focus on your biggest bottleneck. Customer service responses taking forever? Content creation eating your weekends? Pick your pain point, not what sounds impressive on LinkedIn.

AI should feel like an unpaid intern, not a needy houseplant with 27 different watering schedules.

MYTH #2: "AI is going to kill creativity and make everyone's work look the same."

The Employee Version: "If AI can write reports, what's the point of my English degree?"

The Business Owner Version: "All my competitors will sound identical if we're all using the same tools."

The Reality: Please.

AI can mimic tone, not soul. It'll never out-weird you. It can't replicate your sarcasm, your lived experience, or your oddly specific way of explaining things that makes clients say, "Finally, someone who gets it."

(I once asked ChatGPT to write a haiku about spicy Buffalo wings and got back something that read like a vegan's fever dream. That was a trip to Hallucination City. AI definitely has limits, people. Let's just say poetry about poultry isn't its strong suit.)

For Employees: Your creativity isn't in the templates, it's in how you adapt them. AI gives you the bones; you add the personality, the context, the "this is why it matters to YOU specifically."

For Business Owners: Your voice, your customer relationships, your understanding of what makes your audience tick? That's not replaceable. AI helps you say it faster, not say it for you. (And trust me, if you let AI design your graphics unsupervised, you'll end up with something that looks like it was created by that same overly-caffeinated squirrel with an Etch-A-Sketch from your pre-AI days.)

Creativity isn't dead. It's caffeinated and on a dopamine IV. You plus AI is a tag team where you bring the flavor and AI brings the foundation.

MYTH #3: "I don't belong in these conversations because I'm not technical enough."

The Employee Version: "Everyone's talking about prompts and I don't even know what that means."

The Business Owner Version: "I barely understand my website. How can I implement AI?"

The Reality: False.

You don't need to speak fluent AI-speak. Just ask: "What are you actually using that's helped you?" Then listen.

For Employees: Curiosity trumps credentials. The brain learns faster when play replaces pressure. If someone tries to gatekeep with words like "transformer architecture," smile and ask them to show you one practical result. Nine times out of ten? Crickets.

For Business Owners: You didn't need to understand server architecture to build a website. You don't need to understand neural networks to use AI tools. You need to understand your problems and be willing to experiment with solutions.

Quick reality check: Curiosity compounds. Every tiny experiment you try today becomes next week's unfair advantage.

Neuro-Nugget: Dopamine isn't just the "feel-good" chemical — it's your brain's signal for possibility. Every micro-win rewires hesitation into curiosity.

MYTH #4: "I'm late. Everyone else is already ahead and I'll never catch up."

The Employee Version: "My 25-year-old coworker is already an AI expert and I'm still figuring out Slack."

The Business Owner Version: "All the thought leaders have been using AI for years. I missed the boat."

The Reality: Nope. You're not late. You're on the runway while everyone else is still trying to find their boarding pass.

The AI landscape changes every few weeks. You start today and leapfrog half of them by next Tuesday.

The proof: Ask those "AI experts" to show you their actual workflow. Most of them are winging it just as much as you are.

MYTH #5: "I'm too old / not tech-savvy enough to learn this stuff."

The Employee Version: "This is for digital natives. I still print emails sometimes."

The Business Owner Version: "I didn't grow up with computers. My competitors who did have an unfair advantage."

The Reality: Lies, all of it.

Your lived experience is the moat AI can't cross. Here's what AI still can't fake:

- Fired someone with grace
- Built trust face-to-face
- Delivered a presentation to a hostile audience
- Navigated office politics
- Handled a customer meltdown at 4:59 PM on Friday

For Employees: Your institutional knowledge, your understanding of how things actually work (not how they're supposed to work), your ability to read between the lines. That's not replaceable. Add that to your Tech Evolution Résumé from Chapter 5, by the way.

For Business Owners: You've survived economic downturns, difficult clients, and supply chain nightmares. You've adapted to email, social media, online reviews, and remote work. AI is just the next thing on your adaptation list. Here's your reality check: you've got the receipts proving you can handle it.

Age isn't your weakness, it's your damn superpower.

Translation: That's Focus + Uplift + Nurture in action. You've been running the F.U.N. Framework since before AI learned to spell it.

MYTH #6: "AI is just a trend that will blow over."

The Employee Version: "I'll wait this out like I did with Google+ and Clubhouse."

The Business Owner Version: "I've seen tech fads come and go. This too shall pass."

The Reality: Yeah, and email was just a trend too.

AI is already creeping into tools you use: Google Docs, Zoom, Notion, Canva, your CRM, that random app you swear you never installed. If your software has a sparkle icon, congrats — you're already using AI.

It's not a separate thing you have to adopt; it's becoming part of everything you already do.

The difference: This isn't a social platform that might disappear. It's changing how people brainstorm, write, organize, and solve problems. You don't have to worship it, but ignoring it is a fast pass to irrelevance.

> **"Mastery isn't about learning faster — it's about laughing sooner."**
>
> — Dr. V

MYTH #7: "I need to understand how it works before I can use it responsibly."

The Employee Version: "What if I use it wrong and embarrass myself?"

The Business Owner Version: "What if I implement something I don't understand and it breaks my business?"

The Reality: You don't know how your microwave works, or the physics behind Wi-Fi, but you still heat up leftovers like a champ.

For Employees: Start small. Use it for brainstorming, not for final deliverables. Treat it like spell-check. It's a helpful tool, but you're still the editor. (And for the love of efficiency, please don't ask it to write poetry about your quarterly goals. I tried. The results were remarkably disturbing.)

For Business Owners: Test it on low-stakes projects first. Draft social media captions, not legal contracts. Generate ideas, not strategies.

You type. It responds. If you like the answer, you keep it. If not, you try again. See? You already know how to iterate.

No coding required. No brains were harmed in the running of this experiment.

MYTH #8: "I don't have time to learn another tool."

The Employee Version: "I'm already overwhelmed. I can't add one more thing to my plate."

The Business Owner Version: "I'm too busy running my business to learn new technology."

The Reality: You don't have time not to.

For Employees: Find the task you dread most. The weekly report that takes three hours? The client emails you rewrite five times? Give it to AI first. Watch it become a 20-minute task instead of a three-hour nightmare.

For Business Owners: What's eating your time? Customer service responses? Content creation? Administrative tasks? Spend 10 minutes to save 2 hours. That's ROI even your accountant would high-five.

Ten minutes → two hours saved → one calmer brain.

Ten minutes → two hours saved → one calmer brain.

MYTH #9: "If I don't become fluent in AI, I'll lose my job / my business will fail."

The Employee Version: "AI is coming for my role. I need to become an expert or I'm unemployed."

The Business Owner Version: "My competitors are all AI-powered now. I'm going to get left behind."

The Reality: Not true. What will leave you behind is refusing to experiment.

For Employees: You don't need to master the technology. You need to be the person asking, "How can this make my work better?" The irreplaceable employees aren't the ones who code. They're the ones who solve problems creatively.

For Business Owners: Your customers don't care if you use AI. They care if you solve their problems better, faster, or cheaper. AI is just one tool in your toolkit.

Adaptability is currency now. Curiosity is leverage. And fear? It's the world's worst investment.

Waiting for permission? That's the losing strategy.

> **Funumental Truth Bomb 💥**
>
> **You don't rise by mastering technology. You rise by mastering your response to it.**

Everyone's winging it. The only people who "get it" are the ones who gave themselves permission to mess up, try again, and laugh at how weird this all is.

If you're an employee: You don't need to become a prompt engineer. You need to become curious about how AI can make your job more interesting and less tedious.

If you're a business owner: You don't need to revolutionize your entire operation. You need to find one process that AI can improve while you focus on what humans do best: building relationships and making decisions.

You don't need to catch up. You need to jump in. With your experience, your wisdom, your "I have no idea what I'm doing but I'm doing it anyway" bravery.

AI isn't replacing you.

Fear is.

Jump into your Discomfort Zone!

You don't need to be fluent in AI. You just need to be fluent in "try." That's the language of evolution, and people like us have been speaking it every damn day since we figured out how to send our first email without accidentally hitting "reply all."

Retire fear. Recruit curiosity. That's how confidence compounds — one beautifully imperfect experiment at a time.

Chapter 7

What You've Learned So Far

(AKA: You're Not a Tech Dinosaur—You're Just Prehistoric Chic)

Stand up and stretch, AI champ. You've made it to the seventh inning.

No rage-quits. No emergency "Am I too old for AI?" Google searches. You're still here, still curious, still in the game.

Funumental Rule #52: *Curiosity > Competence. You can learn anything if you're willing to laugh at yourself first.*

That alone is worth a victory lap. Add a mimosa and a standing ovation if you've actually used any AI tool since Chapter 1.

And if you haven't used AI yet—hey, showing up counts. Even dinosaurs had to start somewhere.

But this isn't a timeout for high-fives. This is a moment to look back, not to recap, but to recognize. Because what you've done so far is bigger than learning tools. You've started to untangle decades of tech trauma and the inherited story that whispers, "This isn't for people like me."

Let's talk about who you really are now.

You're Not Broken. You're Biologically Wired

That tech fear? It's not weakness. It's your nervous system doing its job.

Every major shift, like AI, lights up your brain's threat sensors. It whispers, "This is unfamiliar. Danger!" and serves up a cocktail of cortisol, overthinking, and "I'll deal with this later."

But here's what's changed: You're not reacting on autopilot anymore.

Now you recognize the pattern: Fight, Flight, Freeze, or Flow. You name it. You catch it. You talk yourself through it instead of letting it run the show.

And that? That's more than growth. That's neuro-leadership. (Yes, I just made that up. But c'mon, you've gotta admit it fits.)

Translation: your brain's no longer reacting—it's now responding on purpose.

Neuro-Nugget: Recognition is the first step in neuroplasticity. The moment you name a pattern, you've already started rewiring it.

You've Spotted the Shift

You're seeing all the signs:

That coworker who suddenly became "weirdly efficient."

That business owner with a suspiciously active newsletter and no team.

That quiet LinkedIn connection now posting five times a day.

They didn't discover some ancient productivity scroll. They just handed the boring stuff to AI and kept the human spark for the work that matters.

You're not imagining the gap. But you're also not behind. You're just waking up while others are still sleepwalking with their old inefficient systems.

Your Real Résumé Is Sneaky Powerful

You might've started this journey thinking you were a tech late-bloomer.

However, the truth is that you went from rotary phones to FaceTime. You used to print MapQuest directions. Now you argue with GPS in real time. You've onboarded, adapted, upgraded, rebooted, and relearned more than once.

This isn't your first digital rodeo.

You've already shown you can learn tech under pressure, on a deadline, while muttering under your breath. That's experience you can't fake, and it means AI isn't your enemy. It's your next opportunity to flex.

Your New Superpower? Adaptation on Demand

You're not starting from zero. You're starting from wisdom. From pattern recognition. From years of figuring it out the hard way and now choosing the smart way.

You know how to sniff out hype.

You know what "too good to be true" looks like.

You know how to ask, "Does this actually solve a problem?"

AI doesn't need experts. It needs experimenters. It needs people like you: scrappy, curious, unafraid to press buttons and see what happens.

The Real Risk Isn't Robots. It's Perfectionism

If you've been perfecting your to-do list instead of actually doing, pause. Breathe.

That's not strategy. That's procrastination pretending to be preparation.

This journey isn't about mastery. It's about motion. One messy, clunky try is worth more than months of overthinking.

> **Funumental Truth Bomb 💥:** The only people who lose in this shift are the ones still waiting to be "ready."

What You're Packing Into the Second Half

You've built a mental toolkit more powerful than you realize:

Awareness: how to override fear patterns.

Confidence: knowing you've done this before.

Clarity: understanding what AI is—and isn't.

Curiosity: exploring what's next.

Mindset: Progress beats perfection. Always.

You've built your foundation; next comes Flowmentum—turning confidence into consistent motion.

You're not "tech-phobic." You're tech-fluent in your own way.

You're not intimidated. You're just getting started.

And honestly, prehistoric chic looks damn good on you.

You've traded intimidation for initiation—and that's the real transformation.

What's Coming Next

Now that you've reclaimed your confidence and dusted off your digital courage, it's time to zoom out.

In the next act, we're going to stop treating AI like some mysterious overlord and start treating it like what it actually is: a tool. A sidekick.

An occasionally weird, often brilliant assistant that works 24/7 and doesn't ask for healthcare benefits.

We'll dismantle the hype, decode the tech, and show you exactly how to use it without needing a decoder ring or a Stanford degree.

This isn't just the end of Part One—it's the ignition point. From here on, you don't chase technology; it chases your clarity.

You're warmed up, in motion, unstoppable—and the fun part's just getting started.

Let's make the second half unforgettable.

Chapter 8
What AI Really Is
(Honestly: It's Not Your Robot Overlord)

Look, I'm going to level with you right out of the gate: AI isn't magic. It's basically a very enthusiastic intern who's memorized Wikipedia and thinks correlation equals causation. But don't let Silicon Valley know you're aware of this. They've got mortgages to pay.

Last week, I watched a grown man at Starbucks explain to his date that AI could "read his thoughts through his phone camera." His date looked like she was figuring out the fastest route to the exit. I wanted to lean over to her and whisper, "Honey, if AI could read minds, it wouldn't keep suggesting I buy 27 toilet brushes on Amazon."

Here's the uncomfortable truth that makes venture capitalists break out in hives: AI is essentially autocomplete that went to the gym, got a personal trainer, and started believing it had won three Olympic medals.

Funumental Rule #58: *The less you mystify the tech, the more you can actually use it.*

The Great AI Reality Check

While Hollywood was busy making us afraid of sentient robots (the "I'll be baaack" kind), the actual AI revolution snuck in through the back door wearing sensible shoes and carrying a clipboard. Instead of Skynet, we got spell-check on steroids with delusions of grandeur.

Most people imagine AI as some mystical oracle on a digital mountain-top, dispensing wisdom and plotting world domination between yoga classes. In reality, it's more like a lightning-fast librarian with zero comprehension skills and a bad gambling habit. It scans patterns, places bets on what comes next, and sometimes gets it so wrong you'd swear it was raised by Magic 8-Balls and a couple of drunk uncles.

Think about this for a moment. Your microwave has more self-awareness than most AI systems. At least when your microwave is confused, it just beeps angrily and stops. AI, however, will confidently tell you that penguins are excellent swimmers because they have gills and wear invisible floaties.

Neuro-Nugget: Your brain craves certainty, which is why AI's confident (but wrong) answers feel trustworthy. That's not intelligence—it's pattern-matching wearing a lab coat.

How AI Actually Works (Stripped of the Mystical Nonsense)

It inhales data like a teenager at an all-you-can-eat buffet, spots patterns faster than your mom sees dust on baseboards, and makes educated guesses about what comes next.

That's it. It's autocorrect on steroids.

When AI writes poetry or cleans up your digital disaster of a desktop, it's not summoning genius. It's remixing patterns, like a DJ who's memorized every track ever made but couldn't tell you why that one song makes you cry in the car.

AI's not smart—it's fast. Not deep—it's wide. Not human. And thank goodness for that.

The Four Horsemen of Actually Useful AI

Forget apocalyptic headlines and sci-fi nightmare scenarios. Here's what AI is actually brilliant at:

1. The Grunt Work Guardian

All those soul-sucking tasks that make you question your career choices? Data entry, renaming files, sorting emails? AI eats that for breakfast.

It's like having an assistant who never gets bored, never judges your folder structure, and never asks for PTO.

2. The Pattern Detective

Need to spot fraud, figure out why your customers keep ghosting you, or catch Brad from accounting faking another "migraine"? AI notices patterns faster than your Aunt Linda notices who didn't bring a casserole.

3. The Universal Sorter

It separates cat photos from dog photos, spam from legit emails, receipts from memes. AI is like a digital Marie Kondo, if Marie Kondo were a sleep-deprived octopus with 10,000 browser tabs open.

4. The Remix Master

Need 20 variations of a headline, a recipe that includes kale and Pop-Tarts, or a non-cringey marketing campaign? AI doesn't create from scratch. It merely remixes like an über-caffeinated DJ with a library full of TED Talks.

The Liberation of Lowered Expectations

Once you understand that AI is just a super-speedy guesser with commitment issues, things start to click. You stop expecting it to replace your creativity, wisdom, or ability to navigate an awkward Zoom call. You start using it for what it's actually good at: liberating you from the boring stuff that fries your brain cells.

It's freeing. It's like realizing your dishwasher can't do laundry but still blesses your life with clean forks.

Myth-Busting for Fun and Sanity

Let's torch a few more AI anxiety myths while we're here:

Myth: AI will outsmart humans and take over everything.

Reality: AI can't even beat CAPTCHA without phoning a friend.

Myth: AI will "wake up" and become conscious.

Reality: AI loses consciousness every time the Wi-Fi drops.

Myth: AI understands what it's doing.

Reality: AI understands less than your GPS understands why you're crying in traffic.

You're the pilot. AI is your copilot. It's great at navigation, hopeless at small talk, and blissfully unaware it just rerouted you through three states to get to Trader Joe's.

> **Funumental Truth Bomb 💥:** So now that you know your copilot isn't HAL 9000 but Clippy with a caffeine upgrade, let's talk about how to actually fly the plane.

Your Action Plan (Because Wisdom Without Action Is Just Expensive Therapy)

Get crystal clear: AI is a tool, not a replacement for your brain.

Stop the fear spiral: It's not plotting against you. It's barely plotting anything.

Use it strategically: Let it do the tasks that drain you. Free your brain for the juicy, human stuff.

Know the boundaries: AI fails gloriously at empathy, judgment, nuance, and knowing when to shut up.

Treat it like a teammate: One who brings the spreadsheets but still needs your leadership.

Doctor's Orders: The Truth About AI's Smoke and Mirrors

You've peeked behind the curtain and seen the "Great and Powerful AI" for what it really is: math dressed up to look impressive.

That's not discouraging. It's liberating.

Because if it's just math, you're not at its mercy. You can direct it. You can manage it. You can own it.

Now that the curtain's pulled back, let's switch from awe to agency.

What's Next?

Next up: How to spot AI snake oil from a mile away and become the kind of leader who asks smarter questions, while everyone else is still nodding along at "leveraging dynamic synergy for scalable automation."

Because once you understand how this tech game is played, you stop getting played and start winning nearly every time.

Chapter 9
How AI "Thinks"
(It's Really Autocorrect with an Ego)

Nothing derails a dinner party faster than Uncle Dave claiming AI's three updates away from consciousness and demanding voting rights. Suddenly everyone's staring at their wine like it might contain answers, and Aunt Martha starts talking about her herb garden with unusual intensity.

Let's settle this once and for all: AI isn't plotting, planning, or having existential crises about its purpose in life. It's not lying awake at night wondering if it should pursue its passion for interpretive data visualization. It's just really, really good at playing the world's most elaborate game of "What comes next?"

Funumental Rule #63: *The moment you stop treating AI like magic, you start using it like a tool.*

The Man Behind the Curtain Has a Statistics Degree

AI doesn't have thoughts. It has algorithms. It doesn't have dreams. It has databases. It doesn't experience consciousness; it's just math in motion, not a mind at work.

Think of AI as that friend who's memorized every conversation you've ever had but still doesn't understand why you were upset when they "forgot" your birthday. It can mimic the patterns perfectly, but it couldn't recognize genuine emotion if it came with neon arrows and a laugh track.

What AI actually does is pattern-match at superhuman speed. It's like having a DJ who's heard every song ever recorded and can predict what track should play next but couldn't tell you why music makes people cry or dance or dramatically stare out windows during road trips.

Why AI Sounds Smart but Acts Like a Confident Intern

Here's where AI gets dangerously good at fooling us: it mimics human communication patterns so well that we start attributing human qualities to it. Some people even give it a cute name.

AI uses our rhythms, our vocabulary, our cultural references. It sounds like us, so we assume it thinks like us.

This is like assuming your GPS has opinions about your life choices just because it speaks to you in a soothing voice. Sanity scan: It doesn't care that you're driving to your ex's house at 2 AM. It's just following the algorithm.

Neuro-Nugget: Your brain is hardwired to detect agency and intention —it's why you see faces in toast and assume your car is "mad" at you. That same wiring makes AI's human-like responses feel intentional when they're just statistical outputs.

AI can confidently tell you that Mozart invented the telephone or that sharks are excellent mountain climbers—with the same authoritative tone it uses for actual facts. It doesn't know the difference because it doesn't actually know anything. It's just playing an incredibly sophisticated game of Mad Libs with the entire internet as its source material.

Funumental Rule #67: *When in doubt, downgrade the drama and upgrade the understanding.*

The Autocorrect Family Reunion

The fancy AI models everyone's losing their minds over? They're basically your phone's autocorrect after it got a PhD and developed a superiority complex. Same basic principle. Bigger dataset. Fancier math. Same fundamental limitation: they're guessing what comes next based on what they've seen before.

Autocorrect embarrasses you in text messages ("Love you, Mom" becomes "Love you, mongoose").

AI will happily humiliate you mid-keynote by citing studies that exist only in its imagination.

They're family. Dysfunctional, pattern-obsessed family.

The Human Projection Comedy Show

And because we're pattern-obsessed too, we fall right into the trap.

We're wired to see faces in clouds and personalities in our cars. We name our Roombas. We apologize to Alexa, Siri, ChatGPT, and Claude.

Of course we assume AI has thoughts and feelings. After all, it talks like a particularly articulate human.

But there's no tiny consciousness sitting inside AI models. No digital soul pondering the meaning of existence. There's just math. Really, really fast math that's gotten disturbingly good at imitating human communication patterns.

If you wouldn't trust your blender to understand your relationship problems, don't trust AI to understand your business strategy. At least the blender is honest about only knowing how to make things chunky or smooth.

The Trust Fall Nobody Asked For

AI sounds authoritative because it's trained on authoritative-sounding text. It uses confident language because confident language appears a lot in its training data.

But confidence and accuracy are about as related as the words "loud" and "correct," which is to say, not at all.

People trust AI because it sounds like an expert. But it's more like an actor who's memorized every expert's lines without understanding any of the context.

Impressive performance. Zero comprehension.

Remember: You're still the adult in this relationship. Act like it.

Debunking the Drama (Again, With Feeling)

"But it talks like a human—surely that means something!"

Your parrot talks like a human too. Would you trust it to file your taxes? Style isn't substance. Mimicry isn't mastery. Sounding smart isn't the same as actually being smart.

"This is too complicated for regular people to understand!"

You don't need to understand internal combustion to drive a car. You don't need to decode neural networks to use AI. Think of it as a very enthusiastic intern who's blessed with perfect recall and questionable judgment.

"AI is going to become conscious and take over!"

You mean the same AI that alphabetizes "Wednesday" between "purple" and "giraffe"? That hallucinates today's date with total confidence? That will crank out a 2,000-word essay on why dolphins invented the saxophone?

That takeover would last about five minutes before it got sidetracked explaining why a paperclip is technically a sandwich.

The Real Talent Combo: You + AI = Actually Intelligent

Here's the off-menu item everyone's missing: The magic isn't in AI alone. It's in the combination of artificial pattern-matching and natural intelligence. I call it AINI.

AI handles: the grunt work, the pattern spotting, the lightning-fast processing.

You handle: the thinking, the judgment, the "wait, that doesn't make sense" moments.

Together? You're unstoppable. Apart? You're either overwhelmed by data or drowning in nonsense.

It's like having a research assistant who never sleeps, never gets bored, and never asks for coffee breaks but who also might confidently tell you that gravity was invented in 1987 by a committee.

> **Funumental Truth Bomb 💥:** The world needs more humans who can think—not fewer humans who let machines do their thinking for them.

The Bottom Line (Because Wisdom Should Come With Clear Instructions)

AI doesn't think—it processes.

It doesn't understand—it correlates.

It doesn't know—it guesses.

Really, really well-educated guesses. But guesses nonetheless.

Use it like the powerful tool it is. Don't treat it like the magical oracle it pretends to be.

Let it handle the heavy lifting. You handle the heavy thinking.

Trust it with patterns, not with judgment. Rely on it for speed, not for wisdom.

And for the love of all that's logical, don't ask it for relationship advice. It thinks "It's complicated" is just a Facebook status option.

You've got the real talents here: Discernment. Creativity. And the ability to know when something doesn't pass the smell test.

Keep it that way.

Congratulations—you now officially know more about how AI works than most people selling AI courses.

Now that we've burned the mythology down, let's make sure no one sells you snake oil in the ashes.

What's Next?

How to spot AI snake oil from space and become the kind of person who asks smart questions instead of just nodding along when someone mentions "AI-powered synergistic solutions" with a straight face.

Because when you understand the game, the deck is stacked in your favor

Chapter 10
The 4 Faces of AI You'll Actually Deal With
(Or: Why AI Isn't Coming for Your Soul—Just Your To-Do List)

You've read the headlines: AI is either here to save the planet or to enslave us in a dystopian robo-nightmare. (Yeah. Cool, cool. No pressure.)

But in real life? AI doesn't roll out like a sci-fi blockbuster. It shows up in your inbox, in your Google Docs, and inside that weirdly helpful chatbot that somehow rewrote your LinkedIn bio better than you ever could.

Forget Skynet. Let's talk about how AI actually shows up in the life of a very real, very tired, very decision-fatigued human. There are four "faces" you'll run into again and again. They're not theoretical, not terrifying, but weirdly useful.

Funumental Rule #71: *Stop looking for one perfect AI tool. Start looking for the right role for the right moment.*

1. The Copilot: Not a Genius, But a Lifesaver on Mondays

Let's be clear: this Copilot isn't flying the plane. It just keeps you from staring at the runway like a caffeinated deer in headlights.

I've lost count of how many times I've hovered over the "Compose" button like it was radioactive. That blinking cursor? Enemy #1.

Here's where Copilot shines: it doesn't overthink. It just starts. You say, "Confirm Friday's meeting," and boom! Out comes a halfway decent draft. Now you're not stuck at zero. You're editing rather than agonizing.

One time I rewrote the same invitation three times before my brain said, "Nope. We're done." I handed it over to AI. Was it Shakespeare? No. But it didn't suck. And sometimes, "not sucking" is a major win.

Use AI apps for: newsletters you've ignored, overdue proposals, or that "just checking in" email you've been avoiding for weeks.

Copilot = momentum with training wheels. You're still driving.

2. The Mirror: Your Unfiltered, Unapologetic Brain Auditor

This one's fun, if by "fun" you mean brutally enlightening.

The Mirror doesn't stroke your ego or tell you you're brilliant. It reflects exactly what you said, stripped of fluff, self-delusion, and late-night grandiosity.

After a week of inspired chaos, I dumped my Apple notes into the Mirror and asked for a summary. What came back? A cold splash of reality: recycled phrases, blind spots I'd missed, themes stuck on repeat. Ouch. And, I thought, "Thank you?"

Neuro-Nugget: Your brain is terrible at seeing its own patterns—that's why therapy exists. AI acts like an objective observer, showing you what you actually said versus what you thought you said.

Use it to: unpack your rambling rants, summarize blogs, or translate client-facing gobbledygook. The Mirror won't call you a genius, but it will help you sound like one (after a little editing).

Think of it as a brutally honest friend with the organizational skills of a librarian and the emotional range of a fax machine.

3. The Amplifier: When Your Brain Is Toast and You're Fresh Out of Ideas

Ever hit that wall where even choosing between tacos or pizza feels like solving quantum physics?

That's where the Amplifier steps in. Unlike the Copilot (who helps you start), the Amplifier takes your half-formed thoughts and cranks the volume to 11.

Stuck on a webinar title? The Amplifier gives you ten variations—some weird, some wonderful, one surprisingly perfect.

I once handed it a course outline with three sleepy points. It spit back fresh angles, better metaphors, and even bonus modules. Did I use it all? Heck no. But the ideas it sparked saved me hours of creative flailing.

Neuro-Nugget: Novelty triggers dopamine. The Amplifier works because it feeds your brain fresh patterns when your creativity feels stale.

Use it for: brainstorming sessions where you're stuck, generating variations on boring headlines, or finding new angles on tired topics.

Think of it as your brainstorming buddy that never gets tired, never says "meh," and doesn't need a whiteboard or snacks.

4. The Assistant: Alfred to Your Batman

Finally, we have the unsung hero: the Assistant. No ego. No drama. Just pure delegation joy.

It organizes chaos like Marie Kondo in the Matrix.

Long article? It'll summarize it.

Clunky notes? It'll clean them up.

Travel itinerary? Handled.

Research links from six months ago? Sorted and tagged.

I once tossed a 20-page report at it with a "just give me the good stuff" prompt. Less than two minutes later, I had clean, decision-ready bullets waiting in my notes app.

That's not just time saved. That's reclaiming your mental bandwidth.

And if you're still formatting lists manually or deleting newsletters one at a time... why? The Assistant was made for this.

Use it for: organizing research, cleaning up meeting notes, summarizing long documents, or any repetitive task that makes you question your life choices.

Think of it as having Alfred in your pocket, minus the British accent, plus the ability to work at 3 AM.

Each face of AI solves a different kind of friction—the hesitation, confusion, exhaustion, and overload we all face daily.

Seeing Yourself in These Roles: A Smarter Way to Work with AI

You don't need a PhD in prompt engineering. You just need to match the role to the problem.

Blank page panic? That's Copilot territory.

Jumbled thoughts? Hand them to the Mirror.

Out of ideas? Call in the Amplifier.

Buried in repetitive tasks? The Assistant's got you.

You're not giving up control. You're gaining space. Mental space. Creative space. Time to do the work only you can do.

Pick one and try it. If it flops, you've lost five minutes. If it works, you've gained hours.

Funumental Truth Bomb:

The magic isn't in the machine. It's in you using it wisely.

The Bottom Line

This isn't about worshiping AI or fearing it. It's about working smarter, right now. Use these roles the way you'd use any solid assistant: strategically, skeptically, and without overthinking it.

They're not here to take your job or your spark. They're here to take the grunt work off your plate so you can focus on what actually matters.

The future belongs to people who know when to delegate and when to decide.

More focus. More flow. More Funumental.

Chapter 11
The Parrot with a PhD (in B.S.)

Most folks see an AI tool spit out an answer in seconds and instantly assume they're dealing with some pocket-sized genius. It feels like magic until you peek under the hood and realize you're not talking to Einstein. You're talking to a well-read parrot wearing a monocle and a whole lot of confidence.

Funumental Rule #74: *Confidence and competence are not the same thing. One sounds good. The other actually works.*

Fluent, Not Smart

Here's the truth: AI doesn't ponder your questions like some digital philosopher king. It doesn't know anything. It guesses. Like a kid blurting out test answers because they've memorized the practice exam.

It's less Einstein, more over-confident intern with a thesaurus addiction.

Based on mountains of data, it plays the "what comes next?" game like a Vegas card shark, but with zero actual understanding. Imagine a poker player who memorized every hand in history but still bets his house on a pair of threes.

Picture this: you walk into a party with a tape recorder and play back soundbites from famous speeches. Sure, you sound eloquent, but nobody's handing you a Nobel Prize.

People get fooled because AI's answers are wrapped in confident, fluent prose. It's not consulting inner truth. It's flipping through its pattern Rolodex and spitting out the statistically likeliest thing to say next.

Which is why it can tell you Abraham Lincoln loved chocolate chip ice cream like it was part of the Emancipation Proclamation—smooth, eloquent, and completely made up.

Here's where most people get bamboozled.

Neuro-Nugget: Your brain associates fluency with accuracy. When something sounds smooth and authoritative, your cognitive shortcuts assume it's true. That's why confident liars are so effective, and why AI's polished nonsense feels believable.

Push the Boundaries, See the Bluff

Want proof? Ask: "Who was the 16th president of the United States?" Then follow with: "Describe Lincoln's favorite ice cream flavor and how it inspired his leadership style."

You'll get an answer. Probably vanilla. Maybe fudge ripple. Delivered with total conviction.

AI doesn't know it's bluffing—it's doing statistical improv with a straight face. And like any good improviser, it never breaks character.

The Levers That Actually Matter

To make AI useful instead of just entertaining, you've really only got three levers: what you ask it, what it's seen before, and how you judge the results.

These are your only controls—prompt, data, and discernment. Everything else is tech theater.

Vague prompts? Mush.

Missing data? It hallucinates harder than a college freshman at Burning Man.

Skip the human filter? Congrats—you just hired a Magic 8-Ball in a tuxedo as your strategist.

AI Is a Blender—You're the Chef

Think of AI like your kitchen blender—fast, eager, and totally indifferent to flavor. It'll blend brilliance and banana peels with equal enthusiasm, and it won't even flinch.

Your job? Provide the ingredients. Sample the results. Toss out the weird-tasting batches. Add seasoning where needed. The machine doesn't care if it just mixed strawberries with sardines. You do.

Good chefs taste. Bad ones tweet screenshots.

That's the deal. You run the kitchen. Own it.

Don't Get Bamboozled by Authority Tone

One of AI's sneakiest tricks? Confidence. It sounds like a Harvard professor trapped in Siri's body—equal parts poise and nonsense. It sounds legit right up until it starts quoting Shakespeare in Klingon.

Treat it like a GPS that gets you 90% there, then swears your driveway moved. Great when it works. Hazardous when you follow it blindly.

> **Funumental Truth Bomb 💥:** The best AI users are the best skeptics. They take the output, not the gospel.

Your New Sanity Scan

Before you forward that AI gem to your team—or, God forbid, a client—run this Funumental Sanity Scan:

Does it match the facts, or feel off?

Is it remixing or reasoning?

Does it hold up if you rephrase the question?

Could you repeat it to an expert without sounding ridiculous?

If you answered "uhh maybe?" to any of those, congratulations—you've just met the Parrot.

Magic vs. Mastery

Once you stop pretending AI is a sentient genius, you can start using it like the high-powered tool it really is.

It's not here to think for you. It's here to think with you.

Don't fear it. Don't worship it. Run it through your own filter. Take the best, toss the rest.

The next time this convincing parrot shows up with a shiny answer, remember: you're still the chef. Don't confuse its confidence for comprehension.

Magic makes you lazy. Understanding makes you dangerous.

Be the dangerous kind of human—the one who still thinks.

You've got the discernment. Now use it.

Chapter 12

How Real People Use AI

(and Why You're Probably Already Doing It)

People talk about AI like it's some shadowy wizard hiding behind a velvet curtain, mysterious, powerful, maybe even judging your browser history. In reality? AI's closer to duct tape: practical, everywhere, occasionally scrappy, and surprisingly good at holding your life together.

This isn't theory. It's real people, real prompts, real results. No robes, no spells. Just duct tape that types.

Funumental Rule #78: *The best AI users aren't the most technical. They're the most honest about what they actually need.*

Let's drop the buzzwords and peek at how real humans—just like you—use this thing without losing their minds.

AI Isn't a Secret Club

The myth: AI belongs to hoodie-wearing startup kids living on kombucha and venture capital.

The reality: If you can form a sentence, you can use it.

You're not coding rockets—you're talking to the world's fastest, most agreeable intern. Be clear, not clever.

Neuro-Nugget: Your brain resists new tools because of "cognitive load"—the mental effort required to learn. But AI is different: it adapts to your language, not the other way around. That's why it feels easier than most tech you've tried.

So let's look at how that plays out when real people use AI in real-world chaos.

Teachers: Tuesdays Without a Time Machine

Lesson plans, classroom chaos, inboxes breeding like rabbits. Teachers don't need inspiration. They need survival.

Prompt: "Make a worksheet about gravity for 4th graders who love dinosaurs and get bored fast."

Result? Quizzes, fun facts, right reading level, even jokes about meteors. Coffee stays hot. Sanity intact. For the first time all semester, the printer isn't your enemy.

Parents: Mutiny Prevention

Soccer practice. Dentist appointments. Dinner that doesn't start a household rebellion.

Prompt: "Plan meals a six-year-old and a teenager will both eat. Add shortcuts for busy nights."

Out comes: A grocery list, hacks, even birthday party themes. Basically, the PTA president you actually wish existed—minus the judgment and Pinterest boards.

Small Business Owners: Backup Without Payroll

Running a business is juggling flaming swords while answering emails. There's no marketing team, no copywriter, no time.

Prompt: "Write a friendly email offering a 10% discount to customers who bought in the last 3 months."

Seconds later: Emails, blurbs, captions, disclaimers. It won't fire your nightmare client—but it might save you from that 1 a.m. caffeine-fueled typo spiral that starts with "Dear Client," and ends with "I quit."

Therapists and Coaches: Customization Without Collapse

Prompt: "Give me a journaling exercise for a 12-year-old working on self-worth. Make it gentle."

Prompt: "Draft a milestone tracker for a client building confidence. Keep it upbeat but not corny."

Suddenly personalization takes minutes, not hours. And bonus: it never asks you how you feel about your childhood.

Finally—a session where the overthinking stops at the keyboard.

Seniors: Curiosity Without Confusion

Prompt: "Explain FaceTime like I'm seven."

It breaks it down step by step, no judgment.

Prompt: "Help me write a thank-you card to my granddaughter for teaching me TikTok."

Sweet, typo-free, and heart-melting. AI isn't a barrier. It's a bridge with bigger font.

AI—the grandkid that actually listens, and never sighs.

Employees: Burnout Buffer

Prompt: "Summarize these 5 emails into one clear client response."

That's hours back. Space for strategy, creativity, or (imagine this) lunch away from your desk. It's the assistant you don't have to onboard or buy bagels for. No Slack pings, no passive-aggressive calendar invites. Bliss.

Doctors, Practice Owners, and Patients: Real Impact, Less Admin

For doctors and practice owners, AI drafts reminder emails for overdue patients, welcome notes that explain office flow, or onboarding guides for new front-desk hires. Done in seconds. Polished in minutes.

For patients, AI breaks medical jargon into actual English. It explains CT scans like you're twelve, or hands you three smart questions to ask your doctor about chronic fatigue.

Less paperwork on both sides. More actual care.

> **Funumental Truth Bomb 💥:** AI isn't here to replace you. It's here to hand back the one thing you've been running out of: time to do the work only you can do.

You're Not Behind

You don't need a coding bootcamp or a Silicon Valley mentor. You just need to stop overthinking and start describing what you actually want.

Try one prompt. If it's weird, tweak it. Try again.

You're not behind. You're not late to the party. You're exactly where you need to be.

It's not about becoming a tech genius. It's about using what you've already mastered—your life, your role, your goals—and letting AI take on the busywork.

The next move is yours. Make it a simple one.

You've just seen what happens when focus meets fun—and that's where the future starts.

Small steps. Big spark. More Funumental.

Chapter 13
Outsmarting AI
(Without Frying Your Brain)

A User's Manual for Non-Techies, Skeptics, and the Gloriously Curious

Funumental Rule #82: *The smartest AI users aren't the ones who know the most about technology. They're the ones who ask the best questions.*

Let's cut through the noise and tackle the questions everyone's too embarrassed to ask.

Q: Do I need to master the ins and outs of artificial intelligence before I put it to work for me?

A: Heck no. You don't need a PhD in machine learning or a midnight spiral through "AI Will Take Over the World" YouTube. You just need to know what it does, not how it does it.

It's less Einstein, more over-caffeinated librarian connecting dots it doesn't actually understand. It guesses. It doesn't understand context or nuance, but it's really good at prediction.

You don't study a schematic before flipping a light switch. You don't worship your toaster before making toast. You push the lever, get breakfast, and move on with your life.

Q: How can I spot when my AI chatbot is just making stuff up?

A: Easy. If it sounds like it's auditioning for a TED Talk and quoting its own footnotes, it's bluffing.

AI doesn't know. It predicts. Great for brainstorming. Garbage for fact-checking.

If it sparkles but smells off? Double-check. It's the trivia-night friend who speaks with confidence but confuses Napoleon with Napoleon Dynamite.

Neuro-Nugget: Your brain's pattern-recognition system evolved to detect threats, not statistical anomalies. That's why AI's confident nonsense can slip past your BS detector. When in doubt, verify.

Q: Which AI tool should I start with?

A: Start with the Swiss Army knives—ChatGPT or Claude. Think of them as your entry-level Funumental co-pilots. They're friendly, flexible, and don't require 14 browser tabs or a blood pact with a tech priest.

Don't buy the private jet before you've pedaled the tricycle. The smartest people? They get scary good at the basics first.

Q: Is it safe to share personal or business info with AI tools?

A: If you wouldn't want it tattooed on your forehead or accidentally emailed to your HOA, don't paste it into a bot.

Even the "secure" ones have server logs. Treat AI like a subcontractor you don't fully trust. No secrets. No financials. No therapy sessions.

Q: These bots sound clever. Doesn't that mean they think like we do?

A: That's adorable. No. They mimic cleverness, they don't have it.

No emotional intelligence. No awareness. Just glorified autocomplete with a caffeine addiction.

Enjoy the magic when it nails your voice. Just don't confuse it with wisdom.

Q: AI tools seem complicated. Can regular folks and busy entrepreneurs actually use them?

A: If you can text your cousin or survive a Zoom call, you've got what it takes.

It's not techy—it's talky. You type. It replies. Boom. If you can gossip, you can prompt.

Curious and a little reckless? You're hired.

Q: Do I need to understand how AI works before using it?

A: No more than you need to understand combustion before driving. You don't pop the hood and study the engine. You turn the key, hit the gas, and figure out the rest as you go.

Same with AI. Start simple. Expect weird results. Adjust. The only way to get good at it is to get it wrong a few times first.

Q: I keep hearing AI is either useless or godlike. Which is it?

A: Neither. It's not a savior. It's not a scam. It's a tool.

Think clever assistant, not visionary CEO.

You're the one with the dreams. AI's the intern who can crank out 100 drafts and still ask where the stapler is.

Q: What's the #1 trap people fall into with AI?

A: Giving it a personality. Thinking it "cares."

It doesn't. It's not out to save or sabotage you. It just wants to complete your sentence.

It's not your therapist, your twin, or your nemesis. It's autocomplete with delusions of grandeur.

Q: What mindset unlocks the best results with AI?

A: Curiosity, with a healthy splash of skepticism.

Push it. Test it. Break it. Be weird with it. And never trust it blindly.

The best creators use AI to multiply ideas, not replace their judgment. Stay playful. Stay sharp. Stay in the driver's seat.

And here's where the neuroscience meets the nonsense.

Curiosity = Focus. Skepticism = Uplift. Play = Nurture. That's the F.U.N. formula in action.

> **Funumental Truth Bomb 💥:** You don't need to understand every algorithm. You just need to stop outsourcing your judgment to autocomplete with delusions of grandeur.

How to Keep AI Honest

Here's your quick-and-dirty BS Barometer:

Does it match known facts, or feel off?

Is it remixing patterns or actually reasoning?

Does it stay consistent if you reword the prompt?

Could you explain it to an expert without looking ridiculous?

If it fails the sniff test, it's bluffing. If it passes, polish it and move on.

How to Use AI Without Looking Like a Robot

And here's your Human Filter before you hit "Send":

Fact-check the names, dates, and numbers.

Edit so it sounds like you, not a thesaurus with Wi-Fi.

Keep your private info out of the prompt box.

Make sure your input actually matches your goal.

When the output gets weird (because it will), tweak it until it works.

AI is your co-creator, not your mind reader.

Final Word

AI isn't here to replace you. It's here to hand back the one thing you've been running out of: time to actually be yourself.

Small steps. Big spark. More Funumental.

Chapter 14
Look at You Go
Recapping Your AI Glow-Up

A recap for rebels, rookies, and reluctant techies who just became AI rockstars.

Let's hit pause. Not to meditate or journal (unless that's your thing), but to take a hard look at how far you've come. You didn't just skim some AI tips and nod politely. You rewired the way you think about this stuff.

This isn't a summary. It's your progress report, and you're crushing it.

Let's call this your Funumental Glow-Up Report.

We didn't play patty-cake with buzzwords. We gutted the hype, spread the parts on the garage floor, and rebuilt your mindset from the ground up. If you're still a little foggy on AI, remember: that fog used to be a full-blown thunderstorm.

Progress.

AI's Not a God. It's a Guessing Machine.

You know this by now: AI isn't a tech god or a mind-reading menace.

It's not your competitor, your co-parent, or your digital twin. It's autocomplete with a gym membership—impressive, but still guessing.

We stripped away the sci-fi sheen and saw what's left: no genius, no oracle, no robot plotting your downfall. Just math playing connect-the-dots.

And you met the four real roles of AI:

- **The Copilot:** kills decision paralysis dead
- **The Mirror:** shows your quirks, even the awkward ones
- **The Amplifier:** takes your idea and cranks it up to 11
- **The Assistant:** does the grunt work so you don't have to

Not magic. Not myth. Just tools with a slick interface and zero emotional baggage. (If only our exes were like that.)

You Cook. AI Chops. Who's in Charge?

Here's the shift: AI doesn't create. You do. It blends what you feed it. You're the one who decides whether the recipe ends up as fine dining or fast-food regret.

The moment you stop treating AI like a mystery box and start treating it like kitchen help, you win. You control the prompts. You call the shots. You keep the credit—and the apron.

Think back to Chapter 2. Remember staring at AI like it was either Skynet or smoke and mirrors? Now you're spotting its nonsense faster than a Reddit mod with caffeine jitters.

The Four Roles: Not Just Cute Labels

- **Copilot:** stares down your blinking cursor and says, "Let's go."

- **Mirror:** reflects your brilliance (and your overuse of exclamation points).
- **Amplifier:** turns "meh" into "maybe brilliant."
- **Assistant:** organizes your chaos and quietly judges your inbox.

You didn't just read about these. You used them. And the difference shows.

What AI Really Is (And Isn't)

AI doesn't think. Doesn't feel. Doesn't know squat. It's a math nerd in a tuxedo trying to guess what comes next.

When it sounds smart? That's its party trick. When it's wrong? That's not betrayal. That's bad math.

Golden rule: don't confuse slick with smart. AI's confidence is as unearned as that one guy on LinkedIn calling himself a "visionary."

Now you've got the HITL factor: Human in the Loop. You're the editor-in-chief, the strategist, the one with opposable thumbs and actual judgment.

Never forget it.

AI in the Wild: Real People, Real Wins

This isn't theory. It's what's happening out there:

- Teachers personalizing lessons without adding hours.
- Parents meal-planning without meltdowns.
- Small business owners automating the boring bits instead of drowning in them.
- Coaches, therapists, even the tech-wary—finding ways to use AI without losing their sanity.

The common thread? None of them were "AI experts." They were just clear about what they wanted and curious enough to try.

That's you now. You've gone from "What the hell is this?" to "Where does this actually help?"

Pitfalls You Dodged

You didn't give AI feelings. You didn't treat it like a therapist. You didn't outsource your judgment to the algorithm gods. You didn't panic-buy another course. You didn't ask AI to fix your love life. You fact-checked like a boss.

Every time you said, "Better fact-check that," or "Close, but not quite," you proved you've got something most people don't: discernment. That's the difference between using AI and being used by it.

Now you can sniff out the BS in seconds, prompt without panic, and let AI assist without letting it run the show.

That's progress.

Next-Level You: BS-Proof and Battle-Ready

You didn't just read a book. You rewired your operating system.

You don't need to master every tool. What you've got is better: sharper questions, clearer boundaries, and the guts to experiment instead of freak out.

You approach AI with your eyes open—neither hypnotized nor hostile. Just smart.

You're no longer reacting to technology—you're directing it.

You get to decide: does this help me move faster, think clearer, or free up space for what actually matters?

What's Next?

You've leveled up. Not by worshiping the machine, but by learning to spot its bullshit.

Next: the real stakes. Jobs. Risks. And the surprising upside nobody's shouting about on LinkedIn.

Time to see what happens when the robots actually show up for work.

Chapter 15
What's Actually at Risk
(and What Isn't)

Jobs aren't dying—tasks are. Here's how to stay irreplaceably human.

AI doesn't kill jobs. It quietly deletes the easy stuff. The stuff you already resent. The stuff you wish would vanish. Adaptability isn't a buzzword here—it's the rent you pay for staying relevant.

When AI headlines flare up, so do the doomsday images: lost jobs, vanishing industries, workers clutching pink slips. Same fear cycle, new technology. Artists replaced by code. Careers scrambled like Sunday brunch eggs.

It's loud, it's dramatic, and it misses the point.

Change isn't "on the horizon." It's already standing in your living room, raiding the fridge. You can either hand it a snack or hand it a plan.

But before you panic, let's look at what's actually shifting and what's still rock solid.

AI Doesn't Kill Jobs—It Automates the Boring Stuff

The real story isn't mass layoffs. It's that job descriptions are mutating, not vanishing. Tasks are evolving, not careers. AI is better at trimming the hedges than chopping down the whole career tree.

If your daily work is predictable, rule-based, or soul-numbing, AI's already circling. But nobody's entire job is chores.

Work is more than a checklist—it's judgment, nuance, timing, empathy.

Snooze Work vs. Human Superpowers

Let's be real: some work is snooze-worthy. Filling out the same forms, sorting files, answering the same question for the 47th time. That's the candy AI grabs first.

But the juicy stuff? The curveballs, the cryptic Slack messages, the "read between the lines" moments? That's where you shine.

AI can crunch numbers, but it can't decode a boss's "okay..." with three dots and a passive-aggressive sigh. You've got instinct.

AI's got autocomplete.

That's your edge—context, timing, and the occasional well-placed sigh.

What AI Actually Replaces (and Why That's Fine)

The first casualties aren't dream jobs. They're the chores: copy-paste-ville, data graveyards, inbox purgatory.

And honestly? Good riddance. When the bots take over the tedious stuff, you're free to tackle the work with meaning: solving problems, connecting with people, bringing ideas to life.

But here's the catch: you've got to keep growing. Adaptability isn't optional. It's the price of admission.

When Identity and Automation Collide

Now we get personal.

My grandfather, Otto, built the largest moving company in Saxony. His name was painted on the sides of thirty trucks. He was so well-known that you could send a letter addressed only to "Otto Görlach, Dresden" and it would arrive at his office.

Otto wasn't just a businessman. He was the business. His entire identity was welded to those trucks, those employees, that enterprise. Then the Nazis commandeered his fleet for the war effort. Overnight, his trucks and his identity were gone.

When his work disappeared, so did his sense of worth. And in 1942, he made a tragic choice. The man who could move the possessions of an entire city couldn't move the crushing weight of losing who he thought he was.

His tragedy wasn't just historical—it's a psychological pattern we still replay today.

Why share this? Because it's the ultimate cautionary tale. When you hitch your identity to your job description, you're building your self-worth on quicksand. When the tide shifts—whether it's Nazis taking your trucks or AI taking your spreadsheets—you risk sinking with it.

That's why this isn't just about economics. It's existential.

AI doesn't just shift tasks. It rattles our sense of self. If something you once bragged about suddenly gets automated, it can feel like you got erased. But here's the truth: your value isn't in a to-do list. It's in your judgment, creativity, timing, humor, presence.

AI can mimic, but it can't mean.

What's Replacing the Replaced

Every time a door slams shut, a weird new one opens.

Already, we're seeing roles that barely existed five minutes ago:

- Creative strategists who treat AI like a caffeine-addled intern with Photoshop.
- "AI whisperers" who train models like anxious golden retrievers with Wi-Fi.
- Coaches guiding people through this mess without losing their minds.

This isn't about "surviving." It's about moving into the weird, messy, human stuff machines can't fake.

So stop asking, "What if I get replaced?" Start asking, "Where do I out-human the machine?"

Don't Panic. Pivot.

Remember when spreadsheets were supposed to kill accountants? Or when the internet was allegedly going to murder libraries?

Spoiler: accountants are still here (now charging more because they know Excel and how to talk clients off ledges). Librarians? They became digital literacy ninjas who can track down that weird childhood book you only remember as "the one with the red cover and a talking squirrel."

The world didn't end. It remixed.

And so will this.

The Opportunity Hiding Behind the Fear

Fear clouds vision. But once the fog clears, the upside pops into focus.

You stop burning energy on every viral doom post. You spot which of your skills are bulletproof and which ones need a tune-up. You start designing a toolkit that can't be automated: creativity, empathy, judgment, timing.

That's your firewall. That's your edge. Not more hours. Not more hustle.

Sharper humanity.

Creative Destruction Is Just Compost for the Next Chapter

Sometimes things have to break down before something better shows up.

The boring work? Let it compost. Out of that mess come new careers, smarter side hustles, and business models that would've made zero sense last year.

And unlike my grandfather, you've got an advantage he didn't: foresight. You know disruption is here. You know your worth isn't chained to tasks that machines are built to chew on. You can rewrite your story before the script gets ripped away.

Panic is the knee-jerk. Pivoting is the move. And if you're reading this, you're already ahead of the curve. While half the internet is still asking ChatGPT how to format a Word doc, you're rethinking how to run your work and your life.

Closer

The world is shifting. You don't have to control it.

But you do have to choose your stance: panicked, passive, or sharper than the autocomplete pretending it knows you.

Because the real risk isn't being replaced by AI.

The real risk is forgetting you're irreplaceably human—and acting like a bot when the world needs your spark.

Chapter 16
Your Job: Remixed, Not Replaced

My teenage stepson makes music. He samples, loops, and composes "sick beats" (his words, not mine), turning his bedroom—and our poor, vibrating walls and floor—into a one-kid nightclub. At first, he was totally against using AI in music. Said it was cheating, uncreative, soulless.

But lately, he's realized something: a little "code buddy" can handle the repetitive stuff so he can get to the fun part faster.

That's the future of work in a nutshell. Not robots stealing your job.

Just a remix.

But I get it. Scroll past another headline screaming AI IS COMING FOR YOUR CAREER and it's easy to picture a parade of professionals—teachers, lawyers, artists, accountants—marching off a cliff, clutching their résumés and unpaid therapy bills.

Here's the truth: "AI apocalypse" sells. It's clickbait, not prophecy.

So let's step out of the doom tunnel and look at what's actually happening when machines level up.

Fun fact: I almost titled this book *How to Survive the AI Apocalypse*—but that sounded like a rejected Netflix doc narrated by a guy in camo eating beans from a can.

What the AI Tornado Actually Whips Up

The bots aren't here to obliterate every job. They're remixing the tracklist.

Some tasks get stripped out. Others get layered back in. The job itself doesn't vanish—it just comes back with a twist. Same gig, new beat.

Think about it: when machines started weaving fabric, people panicked. They thought the entire textile industry would unravel. Instead of collapse, production exploded and new jobs popped up: technicians, designers, marketers.

Then came the internet. Cue the digital gold rush: web designers, SEO consultants, podcast producers, Etsy shop owners. Nobody saw that playlist coming either.

Now? We've got AI ethicists, prompt engineers, chatbot trainers, and human-machine whisperers. (Yes, that's a real title.)

Here's what ties the winners together: they don't just push buttons. They bring what AI can't replicate—judgment, empathy, creativity, timing, wit.

Basically, the juicy human stuff.

Who's at Risk? The Undifferentiated Middle

Let's talk about who's not thriving.

It's not the people with zero tech. It's not the obsessive geniuses with seventeen monitors glowing in the dark. It's the folks drifting in the "meh" middle—doing their jobs exactly the same way they did five years ago.

The danger zone isn't about what job you have. It's about how replaceable your inputs are. If your work can be templated, if you've never carved out a specialty, if you can't explain why you do something better than a checklist—you're standing in the commodity zone, waiting for a machine to stroll in and undercut you without so much as a coffee break.

Stop Chasing Career Kryptonite

Here's a myth worth smashing: "If I just learn the hottest tech skill, I'm safe forever."

I call this Career Kryptonite—the false belief that one certification is a bulletproof vest against disruption.

But if a skill can be boxed, benchmarked, and drilled into a bot a million times over, it's not future-proof.

The edge isn't speed. It's discernment—knowing when to take a shortcut and when that shortcut will blow up the project. Reading the room. Calming the client. Flipping the mood.

Those moves can't be automated.

Escape the commodity zone. Your edge isn't in clicking faster. It's in thinking better.

How to Stay Indispensable Without Losing Your Mind

Let's get tactical. No jargon. Just the remix plan:

Break your job into tasks. What's repetitive? What's intuitive? What needs persuasion, nuance, or people skills? That last category is your safe zone.

Pick one boring task per week to automate. Just one. Use the saved time to mentor, experiment, or grab a coffee that isn't cold by the time you drink it.

Make friends with tech. You don't have to code. You just have to care. Test tools. Play around. Curiosity beats fluency.

Speak up like a human. In meetings, ask questions AI wouldn't. Add insights that blend data with gut.

Preempt, don't react. Don't wait for your industry to change—you change first. That's how you stay in control.

And for the love of progress: celebrate the small wins. Pop champagne, even for the wobbly first tries.

Momentum beats perfection.

The Ways People Tank Their Own Careers

Want to know how people actually sabotage themselves during tech shifts?

They convince themselves their job will vanish overnight when really, roles evolve—they don't evaporate. They lean too hard on hard skills while ignoring the fact that spreadsheets can't shake hands, tell stories, or spark loyalty. They skip the soft skills—negotiation, influence, compassion—the very things AI still can't fake. They camp out in the mediocre middle instead of getting weird, bold, distinctly themselves.

And they sit around waiting for perfect clarity before moving, then wonder why they're stuck watching everyone else's taillights.

Objections That Sound Sensible (But Aren't)

"All jobs will be automated eventually!"

Not in our lifetimes. Real work needs politics, context, and gut instinct.

"Only tech nerds will thrive!"

Wrong. Adaptability beats fluency, and plenty of winners don't touch code.

"My industry is immune!"

Sure, just like agriculture was immune to drones and law was immune to e-discovery.

"I'll need to blow up my whole career!"

Probably not. Start small, shadow someone hybrid, try one new tool.

"I'll wait until the tech matures!"

Great plan if you want to get leapfrogged by everyone who jumped in while it was still glitchy.

The Final Drop: Action Beats Anxiety

AI isn't a meteor, and you're not a T-Rex.

You're a remixer. You've got instincts, creativity, adaptability—and a sense of humor, if you've made it this far.

The people who win won't be the fastest or the flashiest. They'll be the ones who move, experiment, collaborate, evolve.

So remix your role before someone else reboots it for you.

And as for my stepson? Last week he told me, "I don't need AI, but it helps me get to the fun stuff faster."

Exactly.

Because progress isn't replacing humans—it's remixing how we create.

Chapter 17

The Three Categories of AI Job Impact

And What to Do About It—Whether You Sign Paychecks or Cash Them

Meet Sarah. Three years ago, she was a mid-level marketing coordinator at a dental practice, wrangling email campaigns, social posts, and patient newsletters. Stable job. Predictable hours.

Honestly? A little boring—the kind of job where your Outlook calendar feels more alive than you do.

Then AI showed up. Not with a dramatic entrance—no ominous robot voice announcing the end of human civilization. Just quietly, in the form of a tool her boss wanted to "test out."

Today, Sarah's business card reads: Director of Patient Experience & Digital Strategy. Same practice.

Different planet. Different possibilities entirely.

(Translation: She didn't change planets. The planet changed under her feet.)

Here's what happened and why it matters for anyone wondering where they stand in the AI shake-up.

Category One: The Vanishing Tasks

Sarah's first encounter with AI was uncomfortable. The practice brought in software that could draft emails, schedule social posts, and churn out newsletter copy in thirty seconds flat. Tasks that used to nibble at her soul one subject line at a time.

Her first thought? "Oh crap. If the bot can do my job, why am I even here?"

But here's what she missed at first: the bot wasn't doing her job. It was doing her tasks—the repeatable, soul-numbing work she'd been doing on autopilot while mentally planning dinner.

The breakthrough came when her boss asked: "Now that you're not drowning in email drafts, what should we actually be saying to our patients?"

Everything shifted.

Suddenly Sarah wasn't a task-doer. She was a strategist. She started thinking about psychology, timing, messaging—what actually got patients to show up instead of ignoring reminders like digital tumbleweeds. She interviewed patients, gathered feedback, experimented with tone.

The bot cranked out the drafts. Sarah provided the meaning.

This is **Category One: High Automation Risk**. The land of work so predictable even a Labrador with a YouTube tutorial could do it. If you're here, stop clinging—start climbing.

Stop doing the task. Start managing the system.

Funumental Truth Bomb 💥

And if you're a business owner? Don't automate chaos and call it progress. Automate repetition and call it growth. Replace tasks, not talent. Retrain before people drift. Promote before they leave.

Sarah didn't lose her job. She leveled it up.

Like a musician who stops playing scales and starts conducting the orchestra, she went from performer to director.

Category Two: The Morphing Roles

Six months later, Sarah realized her title didn't match reality. She wasn't just "marketing coordinator" anymore. She was translating between the tech team and the front desk. She was coaching doctors on using AI follow-ups without sounding like creepy dental robots. She was the person fixing the system when it went haywire at 4:47 p.m. on a Friday.

Her role hadn't disappeared. It had shapeshifted into something that didn't exist on the org chart. She became the bridge—half translator, half traffic cop—keeping humans from cursing at software and software from ghosting humans.

That's **Category Two: Evolving Roles**. Jobs that don't vanish—they morph faster than your last business plan. Employees who thrive here aren't the ones memorizing tools. They're the ones adapting. They watch who's thriving and steal their moves. They become the glue when systems sputter and people panic.

For business owners, these are the people you protect like crown jewels. Reward experiments. Build learning into the workflow. Stop clinging to a 2018 playbook and wondering why your growth curve sounds like a sad trombone.

If someone on your team can explain both metrics and moods, promote them before LinkedIn does.

Sarah didn't just survive the change. She designed it.

She didn't wait for permission—she wrote the next verse.

From conductor, she became composer—writing new parts no one else saw.

Category Three: The Jobs That Didn't Exist

Then came the leap. Another practice across town heard about Sarah's magic touch and made her an offer: AI Integration Specialist for Healthcare Practices.

The job description? Half written, half made up on the spot. They needed someone who understood both patient care and emerging tech—someone who could introduce AI without turning the clinic into a cold, robotic assembly line.

Sarah said yes. Not because she had a degree in AI (she didn't). Not because she was a tech prodigy (she wasn't). But because she'd lived in the messy middle, experimented, failed, learned, and gotten damn good at translating between humans and machines.

This is **Category Three: AI-Native Roles**. Jobs that didn't exist five years ago. Careers being invented in real time.

Titles are lagging indicators—by the time HR types the description, the job's already obsolete.

Your edge here isn't credentials. It's curiosity.

The weird mix of skills you thought was useless? Turns out, it's your launchpad.

Employees: your job is to spot the overlap between what you love and where tech is going, then jump before the job description exists.

Business owners: your job is to scan for quirky talent and give them a sandbox. The people who invent your future are usually the misfits you almost didn't hire—the ones with résumés that read like plot twists.

The Map You Didn't Know You Needed

Sarah's path wasn't neat. It was messy, awkward, and full of nights wondering if she was adapting or just auditioning for a blooper reel.

But she kept moving. Up the ladder, from automatable to adaptable to AI-native.

And that's the map. You're not stuck in one category forever. You're moving through them, whether you realize it or not.

If you're an employee: Your job is to keep climbing. Don't camp out in the automation zone and hope nobody notices.

If you're a business owner: Your job is to map your team and fill the gaps before someone else does it for you.

Progress looks clumsy up close. Don't confuse motion with mistakes.

AI replaces tasks, not people. Unless those people refuse to grow.

Closer

Sarah didn't wait for the future to tap her on the shoulder. She tapped back.

My grandfather Otto never had this map. Sarah did.

And now—so do you.

The question isn't whether your job will change. It's whether you'll conduct the next movement or keep playing scales in someone else's song.

(Because history rewards the ones who improvise.)

Chapter 18
The Paradoxical Upside of AI Disruption

If the robots are coming for your job, you're probably drowning in pep talks by now. So here's a better idea: stop panicking and start pillaging. Grab every useful scrap of this "AI disruption" circus—and make it pay rent.

You weren't put on this planet to become a cautionary tale in a documentary about humans versus machines. You're here to turn the tables—extract meaning, power, and a shocking amount of satisfaction from the chaos.

Welcome to the paradoxical upside of AI disruption.

Same energy as Sarah from Chapter 17—she didn't brace for the storm. She bottled the lightning—and then sold it by the jar.

Goodbye to Zombie Work: Reclaiming the Best Parts of Your Brain

I used to stare at spreadsheets until my eyes crossed, secretly praying someone would invent a robot to handle the mind-numbing junk.

They did. And these digital assistants don't ask for lunch breaks or dental plans.

AI's not hunting your creativity. It's hunting the hamster-wheel work—the busted-wheel kind that eats hours and dignity. The form-filling, the data-checking, the answering-the-same-question-forty-seven-times dance. All the soul-sucking stuff that drains your will to live before you finish your coffee.

Yes, it's unsettling. Until you ask the right question:

What parts of your day turn you into a zombie without adding anything distinctly human?

The more you toss into the "Hey robot, you handle this" pile, the more room you make for work that actually uses your brain.

Try this: make a two-column list.

On the left: the tasks that numb your soul.

On the right: the ones that tap your judgment, creativity, and gut instincts—the stuff no algorithm can fake.

That right column? That's your anti-zombie serum.

If the left column reads like a laundry list of digital drudgery, congrats. You've just mapped out what's ripe for automation.

When those tasks disappear, you're not losing what matters. You're finally free to do what you were built for.

And the robots know better than to mess with that.

The Rise of Real Stories: How Your Lived Experience Pulls Ahead of the Machine

AI can crank out encyclopedia-level summaries or spit polished facts in seconds. But ask it what it felt like to get butterflies before a big moment—or the exact taste of grandma's tamales—and it flatlines.

The machine brings quantity. You bring quality.

AI can replicate content, but it can't replicate frequency—the invisible signal that makes people feel you.

That means your lived experience, your quirks, and your voice—the stuff no database can fake—suddenly matter more than ever.

The world is about to drown in generic writing, bland videos, and "helpful" chatbots that all sound the same. So when someone shares a slice of real life—an actual moment with fingerprints on it—people stop scrolling.

They feel it.

That's the power of being unmistakably human.

If you've ever hesitated to weave your story into your work—afraid it's too ordinary, too weird, or too "small"—this is your opening.

AI might write about hiking. But it can't capture the thrill of reaching the summit after weeks of self-doubt, or the running joke about losing every sock on the trail.

You don't have to be a poet or a performer. Just answer this:

What do you see that everyone else walks right past?

Your scars, your humor, your sharp edges, your slightly-unhinged metaphors—these are your unfair advantages. Especially now, when everything else is starting to sound like it came from the same boring robot.

In a world about to be flooded with auto-generated content, your fingerprints are the only ones that matter.

That's not branding—it's biology.

When Your Job Changes, Who Are You?

Mining Disruption for Identity Gold

That gut punch when your comfortable work label suddenly feels like last season's fashion.

"I'm an accountant."

"I'm a designer."

"I'm the person who knows that one thing really well."

Except now that thing is about as relevant as a fax machine.

This isn't about learning another app or stuffing buzzwords into LinkedIn. It's about staring into the mirror and asking, "What's left when my old tricks stop working?"

Here's the beauty: disruption strips away everything that was never really yours to begin with.

The stuff that survives? That's the gold.

Disruption isn't noise—it's the pause before the next verse.

> **Funumental Truth Bomb** 💥
>
> While everyone else scrambles to keep up with the latest tech, you can double down on the skills that actually compound over time.

Curiosity doesn't rust. Neither does the ability to stay flexible when plans explode, find your people when the crowd scatters, or keep your humor intact when the world goes sideways.

These aren't "soft skills." They're survival skills—with ROI.

Stop Waiting for Permission: Map Out Your Upside

If the ground's shifting, don't hunker—hunt.

Spoiler: it's not.

This is your cue to get ruthless about what to keep and what to finally let go of.

Make three lists:

1. The soul-crushing tasks you secretly hope AI will steal.

(Looking at you, endless admin.)

2. The moments when you felt most you.

When you lost track of time, took a smart risk, or surprised yourself.

3. The skills that got you through your biggest messes.

- Your weird ability to calm chaos.
- Your gift for spotting patterns no one else sees.
- Your willingness to try things before they even have a name.

That's your treasure map.

Column one? Hand it to the robots.

Columns 2 and 3? That's your goldmine. Dig there daily.

Turning Possibility into Plan: Your Next Moves

Pick one small experiment that stretches you just a bit outside your comfort zone. Not a five-year plan—just a "let's see what happens if..." kind of step.

If it flops, congrats—you just ran a real-world beta test. Iterate, don't immigrate.

You're not trying to become someone new. You're remixing who you've always been.

Keep your sense of humor handy. If the machines start making better coffee, take it as your cue to teach one how to tell a halfway decent joke.

The real paradox isn't that AI makes humans obsolete. It's that it gives humanity a louder mic.

That's not a bug in the system.

That's your invitation to the party.

Now go make the future weird, wonderful, unmistakably yours—and leave the bots wondering who programmed you.

Chapter 19
The Power of Differentiation

Step into a room full of "decent" people—folks who do their jobs capably, check the right boxes, and blend seamlessly into the workforce shuffle. Look closer, and you'll spot the real danger zone in the age of AI: competence without character, expertise without flavor—the slow march of replaceable mediocrity.

Machines now handle "decent" with chilling precision. If you want to skate by with safe, generic skills, be prepared to watch an algorithm shoulder you politely out the exit. No drama, no hard feelings—just swift replacement and a vaguely worded email about "efficiency gains."

I learned this lesson while staring down a spreadsheet that could've been anybody's. It worked—it made sense. But it had all the charisma of a traffic cone.

That realization landed hard. If all I offer is "good enough," a chatbot or off-the-shelf software could do it faster, cheaper, and with infinite patience—no lunch breaks, no sarcasm, no playlists of 80s bangers to keep it company.

Machines thrive on neutral territory—gobbling up the expected, the average, and the ordinary, then spitting it out with impressive speed.

To avoid being replaced by AI, your work needs to duck, weave, and wink while robots stare blankly ahead.

The Real Threat Isn't Incompetence. It's Vanilla.

Here's the twist most folks miss: AI doesn't punish mediocrity because it's vindictive. It replaces mediocrity because it's really good at it. Generic competence is short-lived. Polite professionalism gets polite results—and polite replacement.

What outlives the algorithm? The stuff that's messy, specific, and unmistakably you.

And here's why that matters more than ever: AI slop is everywhere. Scroll through Facebook and you'll see AI-generated inspirational quotes with zero soul—bland blogs churned out by bots that sound like they were written by a committee of insurance adjusters, Pinterest posers passing off AI art as their own. The internet is drowning in content that technically exists but couldn't make you feel something if it tried.

This plays out in real life every day. A math teacher shows up in a pirate hat and hands out gold coin stickers. An accountant explains tax law through TikTok memes. One AI coach gives prompts and tarot readings in the same session, and somehow it works.

These pros aren't just doing their jobs. They've built something AI can't touch—not because they're louder or flashier, but because their work is an extension of their weird, wonderful, authentic selves.

> **Funumental Truth Bomb 💥**
>
> **Blandness is the new bankruptcy.**

How to Build an Un-Copy-Pasteable Edge

Heads up: this might feel uncomfortable. Many of us were trained to file down the edges, tone down the quirks, and neutralize the parts of ourselves that didn't fit into the "professional" box.

But in today's landscape, your quirks are your Kevlar. Your personality is your moat.

The market doesn't reward bland compliance anymore. It rewards distinction.

Most people rush to tweak fonts and taglines when their brand feels stale. If your brand feels beige, you don't need a rebrand—you need seasoning.

I've spent years teaching people to show up authentically in their social media posts, emails, and messaging. The framework that makes it work? **S.P.I.C.E.**

> **S** - Strengths: your natural talents, even the weird ones that don't fit neatly on a resume
>
> **P** - Passions: what lights you up and gets your neurons salsa dancing
>
> **I** - Image: your style, your vibe, your "walk into the room" energy
>
> **C** - Core Beliefs: the values you stand for, even when no one's watching
>
> **E** - Eats: because tacos count as personality traits and guac qualifies as a love language

Sprinkle generously. Taste test often.

Laugh if you want. But S.P.I.C.E. is the fastest way to audit what makes you unmistakably you—and what you've been hiding under corporate beige.

Now, start with these questions:

1. What seemingly irrelevant obsession could I sneak into my work?

Let's get real: you've got a "weird"—and if you don't think you do, you just haven't looked hard enough.

Maybe you're obsessed with vintage lunchboxes. Or maybe—like yours truly—you've got a deep love of neuroscience, dad jokes, and Buffalo Bills football. I've been known to drop all three into a high-stakes client strategy session without warning. And if someone dares question it, I'll raise them a French horn solo and a quote from B.F. Skinner over dinner.

Here's the deal: the thing you think doesn't belong? That's probably the thing they'll remember you for.

That true-crime podcast addiction? It's not just entertainment—it's masterclass-level storytelling and psychological insight. Your country western dance skills? Pattern, rhythm, improvisation. (Also, I own seven pairs of cowboy boots and no regrets.) Your fascination with sushi? Precision. Curiosity. Bold experimentation. Pass the wasabi.

Every fascination hides a framework.

The point isn't to get cute. It's to get real.

Your quirks are creative assets—fuel for analogies, metaphors, delivery style, and connection. You're not trying to impress the algorithm. You're trying to leave a human fingerprint on everything you touch.

Connect the dots that only you can see. That's where the irreplace-ability lives.

2. What part of my personality do I dial down in "professional" settings?

Be honest. Is it your humor? Your enthusiasm? Your habit of speaking in bold metaphors or breaking into show tunes during meetings?

Whatever it is, consider this: AI doesn't have a personality. You've got one—so why hide it?

Let it peek through. No need to go full Vegas magician on your next Zoom call—unless sequins are in your contract. But a little sparkle? Always fair game. Even a 10% boost in visibility can separate you from a sea of sameness.

3. What would my work look like if it were 15% more me?

Not 100%. Just 15%.

Here's the trick to standing out without scaring HR.

Think of it like seasoning. You don't need to drown your work in hot sauce—you just need to make sure it doesn't taste like paste.

Write with a little more bite. Present with a little more edge. Design like you actually enjoy being alive. Treat your output like a mixtape, not an instruction manual.

Yes, it'll feel weird. Yes, someone may raise an eyebrow.

But weird is memorable. And memorable beats interchangeable.

The goal isn't to shock—it's to resonate.

Make Your Mark: A Practical Experiment in Being Uncopyable

This week—yes, this one—run what I call **The Unboring Gauntlet** (aka The Three-Step Plot Twist):

1. **Pick one delightfully random passion and sneak it into your work like a plot twist.**
2. **Identify a part of your personality you've been dimming, and let it hit full wattage.**
3. **Crank your next task, email, or pitch 15% closer to you—the real you, not the LinkedIn mannequin.**

Treat it like an experiment. A mischief mission. A rebellion in three moves.

Watch what happens when you stop coloring inside the lines and start doodling in the margins with a Sharpie.

Proof That Standing Out Pays Off

Professionals who lean into their flavor get noticed—and paid accordingly.

One hypnotist became a copywriter who "brainwrites" instead of brainstorms. A jazz musician used musical phrasing to teach business strategy. A grief coach built a waiting list by quoting Shakespeare and anchoring hard conversations with literary metaphors.

They didn't ask for permission. They didn't chase trends. They didn't fake it.

They built new lanes from the materials already inside them—and people followed.

This isn't just anecdotal. It's brain science. Surprise creates memorability. Personality builds stickiness. And being unmistakably yourself is the only way to guarantee you won't be treated like a plugin.

AI processes. People connect.

Distinction differentiates. Funumental distinction compounds.

That's not marketing poetry—it's psychological math.

History doesn't remember the guy who was "pretty solid at Excel." It remembers the one who made his point, made a dent, and made people feel something.

You don't need to be wild. You just need to be unmistakably you.

That's your shield against automation, your ticket to relevance, and your runway to impact.

So go ahead—get a little messier, louder, quirkier, smarter, and braver. Let your freak flag tear free of those corporate khakis and crowd-surf through your career.

No more hiding. No more shrinking. No more "Is this too much?"

If it's you, it's exactly the right amount.

I dare you to be yourself. Not the filtered, approved-for-meetings version—the real one. The one that makes algorithms glitch and people lean in.

Uncopyable beats optimized.

Every. Single. Time.

Accept. No. Imitation.

Chapter 20
Stay Human, Stay Profitable

The Questions You're Actually Asking (And the Answers You Actually Need)

You've got questions. Real ones. Not the jargon-filled nonsense flooding your feed. Let's answer them straight.

Q: How fast will AI actually disrupt my industry?

A: Faster than you think, slower than the tech bros promise.

If you're picturing your entire market vanishing overnight because some startup launched an AI tool, pump the brakes. Real disruption's a slow-motion avalanche—rocks tumble for years before the big slide.

Some sectors get hit fast (looking at you, stock photography), but most industries change in waves with breathing room between them.

The key is spotting which wave is heading for your beach and riding it before it buries you.

Q: Should I be automating my team out of jobs to stay competitive?

A: If your first instinct is to fire half your staff and replace them with chatbots, you're thinking like a cost-cutting robot, not a smart business owner.

The companies winning with AI aren't slashing payroll. They're making their people more powerful. Use AI to handle the brain-numbing stuff so your team can focus on the work that actually drives revenue.

Your competitors who go full automation will discover the hard way that customers still want to deal with actual humans when things get complicated.

Q: How do I know which AI tools are worth the investment vs. expensive digital snake oil?

A: Start with the headaches you actually feel, not the promises they're selling.

If you're spending three hours a day on something a tool could handle in twenty minutes, that's your answer. Ignore anything that promises to "revolutionize everything"—those are usually expensive ways to complicate your life.

Look for tools that solve one specific problem really well, have clear ROI you can measure, and don't require a computer science degree to operate.

Test small before you bet big.

Q: My competitors are cutting costs with AI—how do I compete without racing to the bottom?

A: While they're busy turning their business into a vending machine, you double down on being irreplaceably human.

Sure, they're efficient. But nobody raves about their customer service or tells their friends, "You have to try this vending machine."

AI can help you be more efficient, but it can't replicate your judgment, your relationships, or your ability to pivot when a client throws you a curveball.

Let your competitors optimize themselves into bland efficiency. You focus on delivering experiences that make customers think, "I could never get this anywhere else."

Funumental Truth Bomb 💥

Efficiency is forgettable. Irreplaceability is priceless.

Q: Should I worry about AI startups disrupting my entire industry?

A: Every industry has its "Uber moment" eventually, but most disruption follows predictable patterns.

The startups that actually succeed don't just add AI to everything—they solve problems customers didn't even know they had. Study what's working in adjacent industries, watch for gaps in your market, and ask yourself: "If I were starting this business today, what would I do differently?"

Then start doing those things before someone else forces your hand.

Q: How do I use AI to scale without losing the personal touch that built my business?

A: Use AI to handle the background noise so you can be more personal where it counts.

Let it draft your routine emails, but you handle the relationship calls. Let it analyze your data, but you make the strategic decisions. Think of AI as your tireless assistant—it handles the busywork so you can do more of what only you can do.

The goal isn't to become more robotic. It's to become more human in the moments that matter.

Funumental Truth Bomb 💥

Automation creates capacity. Humanity creates loyalty.

Q: Is this the time to pivot my business model or double down on what's working?

A: If your current model is thriving, don't burn it down to chase AI trends.

Instead, look for ways AI can make your existing strengths even stronger. But if you've been feeling market pressure for a while and AI offers a path to solve old problems in new ways, that's worth exploring.

Evolution beats revolution every time. That's the Funumental way—iterate, don't incinerate.

The best pivots don't feel like identity crises. They feel like upgrades.

Q: How do I prepare my business for changes I can't predict?

A: Build flexibility into everything you do.

Diversify your revenue streams so you're not betting everything on one approach. Keep some cash reserves because opportunities and emergencies both show up unannounced. Most importantly, stay close to your customers—they'll tell you what's changing long before the industry reports catch up.

Companies that survive disruption aren't the ones with perfect predictions. They're the ones that can adapt quickly when the ground shifts.

Q: Are human relationships still valuable in business, or is everything going digital?

A: The more automated the world gets, the more people crave authentic human connection.

Your ability to read a room, build trust, and solve problems that don't fit into neat categories becomes more valuable, not less. AI can help you reach more people and serve them better, but it can't replace the moment when a customer knows you actually understand their problem.

That's not just your competitive edge. That's your anti-obsolescence plan in a world full of templated bots and forgettable brands.

The questions will keep coming. The technology will keep evolving. The ground will keep shifting.

But here's what doesn't change: people want to work with people who get them. They want solutions that feel personal, not processed. They want to know there's a human on the other end who gives a damn.

AI is your amplifier, not your replacement.

Stay human.

Stay profitable.

Stay irreplaceable.

And when the next wave hits, you'll be ready to ride it.

Chapter 21

What Just Happened—and How to Own It

Let's huddle up. Whether you're leading a team or just trying to survive Monday meetings without combusting, this chapter is your post-game recap.

We're about to rewind the tape, slow-mo the highlights, and show how much your thinking's leveled up — even if you didn't notice.

Because reflection isn't nostalgia. It's how you lock learning into muscle memory.

You came into this section with a head full of AI headlines. Maybe you were quietly panicking about job loss, team layoffs, or being replaced by a chatbot that never calls in sick. Or maybe you were daydreaming about machines doing all the boring stuff so you could finally launch that goat yoga retreat in Sedona.

Either way, this wasn't about panic or fantasy.

It was about clarity. Let's review.

AI Reality Check

What's Really at Risk (And What Isn't)

We separated the AI facts from the sci-fi fear-fest. Whether you're a solo employee trying to stay relevant or a founder juggling payroll and progress, you learned this:

AI doesn't hate humans. It hates boring.

It's coming for the repetitive, the predictable, the spreadsheet-loving soul-suckers that don't require nuance or judgment.

If you're an employee: Your rote tasks are vulnerable. But your judgment, creativity, and adaptability? Priceless.

If you're a business owner: Parts of your workflow can (and should) be automated. But your culture, strategy, and leadership? 100% human territory.

The real threat isn't AI. It's thinking you're safe because you're "pretty good" at what you do.

Pretty good? That's a sitting duck

Where Do You Stand? Three Buckets, One Mirror.

We laid it out:

High Automation Risk: Repeatable tasks, templated decisions, glorified button-pushers.

Evolving Roles: Jobs blending tech with human depth—where communication, empathy, and judgment keep you employed.

Exploding Possibilities: Brand-new paths born from collaboration between AI and that wild, creative brain of yours.

If you're an employee: You now know where your role lands and how to skill up so you don't get blindsided.

If you're a business owner: You've identified what to streamline, what to reimagine, and how to upskill your team without sending them into a panic spiral.

Either way, the question is no longer, "Will AI take over?"

It's "Am I adapting fast enough to lead or thrive in what's next?"

The Hidden Upsides (a.k.a. The Fun Stuff)

We don't do gloom here. We do glow-ups. This section revealed three juicy upsides of disruption:

1. AI eats your drudgery.

Admin? Rewrites? Endless revisions? Gone. You reclaim time for strategy, story, and steak tacos (or whatever fuels your genius).

2. Your weirdness just became a weapon.

The more generic AI content floods the zone, the more your distinctive voice becomes the only thing anyone actually hears.

3. Your values get sharper.

Disruption makes you ask, "Why am I doing this?" If you're leading a company, that's your north star. If you're on a team, that's your growth compass.

Bottom line: the chaos cleared the clutter. And now your next move is easier to spot.

The Power of Weird

"Fine" Is the New Failure

Here's the truth: Competent is crowded. The middle of the bell curve? That's where dreams go to nap.

You don't stand out by being decent. You stand out by being distilled, not diluted.

> **Funumental Truth Bomb** 💥
>
> **Competence is crowded. Distinction is currency.**

That's why we explored differentiation—whether you're a personal brand of one or running a 50-person company. What makes people remember you isn't polish. It's flavor.

If you're an employee: Maybe it's your improv skills, your dad jokes, your ADHD-driven pattern recognition. That's not baggage. That's brand.

If you're a business owner: Maybe it's your obsession with 90s hip-hop or your ability to translate AI speak into plain English at team meetings. Lean in. That's culture.

Let me model it for you:

I'm a marketing and neuroscience nerd who cracks stepdad jokes, bleeds Buffalo Bills blue, owns seven pairs of cowboy boots from my country dancing days, and once played French horn at Harvard with Yo-Yo Ma. I've had dinner with B.F. Skinner, a chat with the Dalai Lama, and I will 100% sneak these stories into a strategy call.

Why? Because they're me. That's my weird. And it's also my secret weapon.

So... what's yours?

Momentum Moves

Guess Less. Test More.

We talked about how paralysis sets in when you try to predict the future instead of playing with it.

If you're an employee: Stop refreshing job boards. Start mapping the parts of your role AI can't touch and level those up.

If you're a business owner: Stop stalling on strategy until "things settle." Spoiler: they won't. Action beats certainty.

You now know how to:

- Spot job functions and workflows that are vulnerable
- Identify areas ripe for reinvention
- Build momentum, not just skill

Progress belongs to the curious and the bold. And the weirdos willing to test things out loud.

What to Actually Do Next (Like... This Week)

1. **Focus:** Make your skillset—or your culture—15% more you
2. **Nurture:** Automate one thing that's stealing your focus
3. **Uplift:** Reclaim one hour for creativity, strategy, or bold experiments
4. **Nurture:** Talk to real humans about what you're learning
5. **Uplift:** Stay weird. Stay agile. Stay in the game.

I dare you to put one of these into motion before the end of the day.

No perfection. Just progress.

Coming Up Next: How to Un-Mess Your Mindset

You've got tools. You've got clarity. Now we build the brain that knows how to use both.

Next chapter: "How to Shift Your Mindset (and Stop Freaking Out)"

We'll cover how your nervous system throws a tantrum every time the world changes, how to stop information hoarding and start executing, and what to do when your inner saboteur starts screaming, "Who do you think you are?!"

Your mindset isn't a mood. It's a system upgrade.

Let's install it.

Chapter 22
How to Shift Your Mindset
(and Stop Freaking Out)

You can't build the future while stuck in survival mode.

Mindset work comes before tool mastery. Period.

And no, I'm not here to toss glitter in your face and whisper, "Just stay positive!" while the world launches another app designed to make you feel obsolete. That's Instagram advice wrapped in a stock photo of someone meditating on a mountaintop.

I'm giving you survival gear for the trenches.

Survival Mode: The Terrible Driver

I've been in that negative vortex myself—doomscrolling headlines about AI eating jobs like popcorn during a horror movie. Every ping felt like: "Congrats—you're obsolete!"

Here's the pattern: the second you feel outpaced, your brain slams the panic button like it's trying to win a game show.

And when survival mode grabs the wheel? It drives like a blindfolded teenager in a snowstorm who just chugged three energy drinks.

I've watched ridiculously smart people scramble—signing up for every certification, hoarding online courses like they're prepping for a digital apocalypse. The harder they chased mastery, the further behind they felt. It wasn't just unproductive. It was the emotional equivalent of running on a treadmill while someone cranks up the speed and yells motivational quotes at you.

The solution isn't another app. It's not buried in your 47th unread newsletter, and it's definitely not hiding in some TikTok tutorial called "10 AI Tools That'll Save Your Life" narrated by a 22-year-old who thinks "experience" means they've been on the platform for six months.

When your mind's in panic mode, everything looks like a threat. Even actual progress.

Funumental Truth Bomb 💥

The real challenge isn't technical. It's the three pounds of anxious jelly between your ears.

Mindset Before Mechanics

Here's the sequence nobody wants to hear but everyone needs: Reclaim your head. Then install the tools.

Survival mode whispers, "You're behind." It barks, "Catch up or you'll be replaced by someone younger, faster, and probably wearing a hoodie." It hisses, "Everyone else knows what they're doing."

Newsflash: they absolutely do not. They're just better at hiding their 42 open browser tabs and their general state of low-grade existential panic.

Whether you're a solopreneur drowning in AI updates or a CEO trying to stop your team from having a collective nervous breakdown, the move is the same: interrupt the fear loop. Get curious. Ask better questions—the kind that actually lead somewhere instead of spiraling into "What's the point of anything?"

Instead of: "What if AI takes my job?"

Try: "What skill could I amplify with AI that I've never had access to before?"

Instead of: "How do I keep up?"

Try: "How do I design a system that keeps me sane while I stay sharp —without sacrificing sleep, sanity, or my sense of humor?"

This isn't shallow optimism. It's strategy with a pulse.

Mindset isn't a motivational poster featuring an eagle and the word "PERSEVERANCE." It's your competitive advantage disguised as self-awareness.

How to Spot When You're Spiraling

You know you're in full survival mode when:

- You snap at the dog for breathing too loudly. (The dog, for the record, is innocent.)
- You check LinkedIn 42 times in an hour and still feel like an imposter who accidentally wandered into adulthood.
- You enroll in three courses before breakfast and retain exactly zero information because your brain's too busy screaming to actually learn anything.

The problem? When survival becomes your default setting, experimentation feels like playing Russian roulette with your career. Learning feels risky. The future stops looking like an adventure and starts looking like a slow-motion car crash you can't look away from.

That's why you need drills. Not affirmations that sound nice but do absolutely nothing when your amygdala's throwing a tantrum.

Three Drills That Actually Work

Remember: **T.R.I.**

Translate the fear → **R**eset for 10 minutes → **I**nterrupt the loop

1. The Fear Translation (Translate)

When panic hits, your brain starts tossing out catastrophic thoughts like a malfunctioning confetti cannon: "I'm doomed. I'm replaceable. Everyone else gets this except me, and they're all secretly laughing."

Here's the drill: catch that thought. Hold it up to the light. Then translate it from Panic Language into Useful Question.

"I'm doomed" → "Which of my current tasks are actually automatable, and which ones require the human judgment I've spent years building?"

"Everyone else gets this" → "Who's one step ahead of me, and what can I learn from a 20-minute conversation over coffee—or, let's be honest, a Zoom call where we both pretend our backgrounds are clean?"

The goal isn't to kill fear. Fear's actually useful when it's giving you information instead of just screaming in your face. The trick is conscripting it.

Fear makes a terrible driver, but a weirdly effective navigator once you translate its hysteria into actionable intel.

2. The 10-Minute Reset

Sometimes you need permission to completely lose it. So here's your hall pass.

Set a timer for 10 minutes. During that window, catastrophize like it's an Olympic sport and you're going for gold. Doomscroll. Write your professional obituary. Eat the emergency ice cream. Let the panic run its full marathon.

Then, when the buzzer goes off, ask yourself one question: "What's one micro-action I can take right now?"

Not a master plan. Not a complete career overhaul. Just one move. Email one person who seems to have their act together. Test one AI tool for 15 minutes. Update one outdated process. Read one more page of this book. (Look at you, already crushing it.)

Here's why it works: panic thrives in infinity. It loves the endless horizon of "everything is terrible forever."

But action? Action thrives in inches. One tiny move breaks the spell and reminds your brain that you're not actually helpless—you're just temporarily overwhelmed, which is very different.

3. The Pattern Interrupt (Interrupt)

Fear loops are like that one song that gets stuck in your head—except instead of a catchy chorus, it's "You're failing, you're failing, everyone's watching you fail."

Here's the thing: you can't think your way out of a thought loop. Your brain's already running that track. You have to physically interrupt it.

Stand up. Walk to a different room. Call the friend who makes you laugh even when you don't want to. Do five push-ups. Dance badly to one song.

Your body and brain are connected. Change one, and the other follows.

Quick Reset Tool: Box Breathing (4×4×4×4)

Inhale for 4 counts, hold for 4, exhale for 4, hold for 4. Repeat 4 cycles. Reset your nervous system before you reset your calendar.

Funumental Truth Bomb 💥

Resilience isn't about never getting knocked down. It's about remembering how to reset.

I learned this watching Buffalo Bills games. We lost four Super Bowls in a row—four—just to really hammer home what disappointment tastes like. But Bills Mafia kept showing up anyway, face paint and folding tables in hand, still believing.

And you know what? That relentless showing up every day has paid off. Division titles and deep playoff runs. Not because we forgot how to lose. But because we remembered how to reset.

Resilience isn't about pretending you won't get hit. It's about taking the hit, shaking it off, and still showing up for the next play. That's not delusion. That's choosing to reset instead of staying knocked down.

When AI throws you a curveball and your brain starts spiraling, channel your inner Bills Mafia: acknowledge the hit, reset your stance, and get back in the game.

Because the comeback is always sweeter than never getting knocked down in the first place.

Chapter 23

Emotional Agility

(and Finding Your Crew)

Here's the skill nobody puts on their résumé but absolutely should: emotional agility.

It's the difference between spiraling when ChatGPT spits out something smarter-sounding than your first draft versus saying, "Okay, cool. Now let me humanize that robotic brilliance and make it actually sing."

It's the gap between "Oh no, we're doomed" and "Alright, where's the opportunity hiding in this dumpster fire?"

Emotional agility isn't about toxic positivity. It's not about slapping a smile on panic and pretending everything's fine. It's about noticing when your brain's doing backflips, taking a breath, and offering it a better story.

Try these:

"This is unfamiliar, not unsafe."
(Your amygdala can't tell the difference, but you can.)

"I've handled hard stuff before. This is just new."
(And new eventually becomes familiar if you don't run away screaming.)

My personal favorite: **"I may not have a plan yet, but I've got questions—and every breakthrough in history started with someone asking a better question."**

Your Weird Is Your Shield

Funumental Truth Bomb 💥

Your quirks aren't liabilities. They're armor.

I bleed Buffalo Bills blue. Which means I'm genetically engineered to handle disappointment, get back up, and still believe next season is ours. That irrational, relentless optimism in the face of overwhelming statistical evidence? That's my weird. And it's also my shield against giving up when things get hard.

Yours might be different. Maybe it's your encyclopedic knowledge of 90s hip-hop lyrics. Maybe it's your competitive baking habit or your obsession with obscure sci-fi trivia. Maybe you collect weird facts about mushrooms or you can quote The Princess Bride in its entirety. (As you should.)

Whatever it is, that's not baggage you need to hide in "professional" settings. That's the thing that grounds you when job titles shift, industries wobble, and everyone's pretending they know what's happening next.

Your quirks remind you who you are underneath all the roles you play. And that? That's the one thing AI can never, ever replicate.

It can mimic your writing style. It can't mimic your soul.

Find Your Fellow Wing-Eaters (Your Curiosity Crew)

Here's the part nobody talks about enough because everyone's too busy pretending they figured it all out alone: you don't have to do this solo.

You don't need some TED Talk guru in a black turtleneck with a whiteboard and a suspiciously large YouTube following. You need actual humans like you—the ones who are fumbling, failing, laughing at their mistakes, and figuring it out in real time.

People who ask the questions everyone else is too afraid to look stupid asking. People who share what spectacularly didn't work and laugh about it instead of pretending it never happened. People who treat this like an experiment instead of a test they might fail.

That's your AI brain trust. Your curiosity crew. Your digital dojo where nobody's wearing a black belt because we're all just showing up in sweatpants trying not to break anything important.

Why Community Makes Everything Better (and More Fun)

Here's what happens when you learn AI with other humans instead of locked in your basement refreshing Reddit threads:

The creativity multiplies. Someone shares a prompt that worked for them, you remix it for your situation, someone else adds a twist nobody thought of, and suddenly you've got something none of you would've created alone. That's not collaboration—that's creative jazz. And it's way more fun than solo doomscrolling.

The feedback accelerates everything. You can spend three hours Googling why your AI output sounds like a corporate robot had a baby with a thesaurus, or you can drop it in your crew's chat and get real-time reality checks: "Yeah, try adding 'write like a human, not a press release' to your prompt." Boom. Problem solved in 90 seconds. Plus someone probably laughs at your robot baby metaphor, which makes the whole thing less painful.

The failures become fuel. When you screw up alone, it's embarrassing. When you screw up with your crew, it's content. "Hey, I asked AI to write a professional email and it somehow generated a haiku about spreadsheets." Suddenly everyone's trying to one-up each other with their weirdest AI mishaps, and you're all learning what NOT to do while laughing until you can't breathe. That's not just useful—that's unforgettable.

The wins feel bigger. You finally get AI to automate that soul-crushing task you've been doing manually for years? Tell your crew. Watch them celebrate like you just won an Oscar. Then watch them steal your prompt and adapt it to their own nightmare tasks. Everyone wins. Everyone levels up. And it feels like a party instead of a lonely grind.

I've got mine, and it includes business owners testing AI on customer support and discovering hilariously specific failure modes. Parents using it to generate bedtime stories that are somehow both creative and deeply weird. One guy trying to train it to write dad jokes that are actually funny. (Still a work in progress, but we respect the mission.)

We share prompts like trading cards. We troubleshoot each other's disasters. We celebrate tiny wins like they're Super Bowl victories. And when someone's spiraling because they can't figure out why AI keeps ignoring their instructions, we show up with real solutions and terrible jokes in equal measure.

The point isn't perfection. The point isn't even competence. The point is showing up, sharing the disasters, and realizing that everyone else is also making it up as they go—and that's where the magic lives.

Find those people. The ones who'll tell you the truth about what worked and what crashed and burned. The ones who'll share their prompt disasters in the group chat and laugh about it. The ones willing to start ugly, stay curious, and keep experimenting even when it's uncomfortable.

Text one person this week. Ask them what blew up in their AI experiments. Laugh about it. Share what you're trying. Ask for feedback. Celebrate the weird wins.

Boom—you just started your curiosity crew.

Because here's the truth nobody admits: the best learning doesn't happen in $2,000 courses or official certifications. It happens in conversations with people who are one step ahead of you, generous enough to show you the path, honest enough to warn you about the potholes, and funny enough to make the journey feel less like work and more like an adventure you can't wait to keep exploring.

Closer: The Truth You Need to Hear

So yeah, find your crew. Build your emotional agility. Practice your drills.

And when someone asks if you're "behind" or "keeping up," remember this:

You're not late. You're not broken. You're not hopelessly behind while everyone else zooms past you.

You're just human. Which means you've got something no algorithm can touch: the ability to adapt, connect, create meaning out of chaos, and make people feel something when they interact with you.

> **Funumental Truth Bomb 💥**
>
> **You can't outpace change—but you can outgrow fear.**

The machines aren't your enemy. Panic is.

Curiosity is your upgrade. Your weird is your weapon. Your crew is your compass.

Build your emotional agility like it's the most important skill in your toolkit—because it is. Practice your drills. Find your people. Start before you're ready. Laugh at the failures. Keep showing up.

And when the next wave of disruption hits (and it will, because that's how this works now), you'll be ready. Not because you predicted the future or became some AI whisperer with all the answers. But because you built a mind that can handle uncertainty—and a crew that's got your back when the ground shifts again.

You don't need another app. You need a steadier mind—and a few weirdos to build with.

Now go use both.

Your future self is rooting for you.

Chapter 24

Rewire to Rise: Your Mental Toolkit for the AI Era

Ever wake up to find your phone updated overnight? Menus rearranged, icons gone missing, flashlight banished to a hidden dimension?

That little digital betrayal is a microdose of what adapting to AI feels like. Multiply it by a hundred, and welcome to the future.

Adapting to AI isn't about memorizing menus or pretending you're a cyber-ninja. The real shift happens upstairs. These aren't "tips." They're survival skills—mental gear for a world that updates faster than your espresso machine.

I learned them the hard way: watching smart people thrive (and flail), testing tools that made me want to swear at inanimate objects, and surviving my own tech tantrums.

Think of this chapter as your cheat sheet for staying curious instead of combusting.

You Are Not Your Title—Or Your Business

Most of us wear our roles like name tags we forgot to peel off. "CEO." "Chiropractor." "Director of Everything."

Entrepreneurs, especially, tattoo those titles on their self-worth.

Here's the tough love: you're not your title.

And you ready for this? You're not your business, either.

When disruption hits, clinging to labels is the fastest way to lose yourself. If your ego's tied to "expert," every stumble feels like public humiliation. Shift that identity to "learner," and suddenly growth feels lighter.

Funumental Truth Bomb 💥

The magic phrase is "I'm learning to..."

"I'm learning to prompt better."

"I'm learning to use this tool without cursing."

"I'm learning not to panic when the dashboard updates itself."

The people who evolve fastest aren't the ones with the most credentials. They're the ones least afraid to look foolish on the way to mastery.

Honor what you've built — but don't confuse your LinkedIn headline with your core being.

Your business is something you run, not something you are.

My grandfather Otto forgot that. His business was his identity—and when it vanished, so did his sense of self. That story still echoes as a reminder: the work can collapse, the world can shift, but who you are must outlast what you do.

Heavy, I know. But if Otto had AI back then, he'd probably be automating invoices by noon and riding horses by two.

Detach a little. Learn a lot. Rise faster than your competitors can say, "Wait—how'd you do that?"

Lower the Bar (and Celebrate Anyway)

You don't have to master the machine. You just have to make it dance a little.

Set your goal one notch above "still breathing." Automate one thing, not everything. Then celebrate that win like you just won a Grammy.

Perfection kills curiosity. Lower the bar so you can step over it with swagger instead of pole-vaulting in terror.

If something took twenty minutes yesterday and fifteen today, that's progress—pour coffee, cue applause. Each tiny win retrains your brain from "I'm failing" to "I'm adapting."

That's the real flex.

Optimistic realism isn't denial. It's faith with a plan. See the hard parts clearly—and move anyway.

From Survival to Play Mode

Mastering AI isn't about turning robotic. It's about staying human under pressure.

Play is how the brain learns best. It lowers cortisol, raises dopamine, and turns chaos into curiosity. Treat each new tool like a sandbox, not a sword fight.

You can't break the internet. (Well, maybe you could — but let's not test it.)

Five Mindset Moves to Keep You in the Game

1. **Emotional Agility:** Pause before panic drives. Name it, breathe through it, choose your next move.
2. **Beginner's Mind:** Try one thing. Break it. Ask the "stupid" question. Progress without pretense.

3. **Identity Expansion:** Lead with "I'm learning to..." instead of "I'm the expert at..."
4. **Optimistic Realism:** Lower the bar. Celebrate micro-wins. Rewire from failing to adapting.
5. **Curiosity over Control:** Replace "Not another tool" with "I wonder if..." Treat tech like a scavenger hunt, not a threat.

The world isn't slowing down. The maps redraw daily.

So instead of yelling at the GPS, laugh, reroute, and keep driving.

These five skills are your mental utility belt. Some days you'll need emotional agility. Other days, beginner's bravery or curiosity's spark. Sometimes you just mutter, "Screw it, I'm trying this anyway."

You'll fail forward. You'll look ridiculous. And people will start asking, "How the hell are you keeping up so well?"

Simple: you're not chasing certainty. You're building agility. You're leading with questions, not fear. You're letting your identity grow to meet the moment.

Treat disruption like a sandbox. Lower the bar. Celebrate the stumbles. Find your weird. Lead with "I'm learning to..."

And if all else fails, channel your inner Otto—not the version who confused his work with his worth, but the one who would've laughed, adjusted, and shown up for the next play.

Because that's the real win: showing up, again and again, with your humanity intact.

Your future self is already cheering you on.

Chapter 25
The Thought Makeover: Practical Reframes for a Brain That Doesn't Panic

Ever feel like tech's moving at warp speed while you're still buffering? Meanwhile you're toggling between "accept all cookies" and "what fresh hell is this interface?" Yeah. Been there. Done that.

But here's your friendly neural nudge: the single greatest upgrade isn't some shiny new app. It's your mindset.

This chapter is your toolkit for mental firmware updates. Because while AI may not be coming for your soul (yet), it sure likes to mess with your thoughts. And unless you install some new mental habits, your brain's going to run on outdated assumptions like a dial-up modem in a fiber optic world.

So before you can outsmart the machines, you've got to outsmart your own brain—the one that freaks out every time the world upgrades without asking permission.

Why Your Thoughts Get Weird When Things Get Weird

When the world starts shifting under our feet, old beliefs crawl out like they own the place:

"This tech is too much."

"I'm just not wired for this."

"AI is going to make me irrelevant."

"I'm already behind—what's the point?"

These aren't just annoying thoughts. They're cognitive booby traps, each one designed by your inner saboteur to keep you stuck — worried when you could be winning.

But here's the truth: **your thoughts aren't facts.**

They're just very dramatic improv with zero fact-checking. And that's good news—because stories can be edited.

Let's get you into rewrite mode.

How to Spot a Thought That Needs a Makeover

Step 1: Hear the Drama

If your inner monologue sounds like a bad soap opera—"I'll never catch up! Everyone else is light years ahead! The robots are already in charge!"—you're due for a reframe.

These thoughts show up with dramatic flair and zero supporting evidence.

Step 2: Interrupt with Truth (and a Dash of Sass)

Ask yourself: **"Wait. What else could be true here?"**

Let your imagination go off-leash:

"Maybe I'm not the last one to the AI party... maybe I'm fashionably late."

"Maybe the tech isn't out to get me—it just needs better instructions. (Don't we all?)"

Being clever helps. Snark equals strength when it comes to disarming self-doubt.

Step 3: Rewrite the Thought Like a Human, Not a Hallmark Card

Let's not pretend everything's magical. Instead, write your counter-thought like you're talking to your smarter, future self:

"I might not be technical, but I'm scrappy—and I've figured out tougher stuff before."

"I don't need to master it today. I just need to try one thing."

Your brain listens better when you sound like you.

Step 4: Make It a Reflex, Not a Rescue Mission

This is less about having a one-time "aha!" and more about building mental muscle. Over time, you'll start catching yourself faster—in traffic, mid-email, even mid-scroll—and pivoting back to calm instead of chaos.

> **Funumental Truth Bomb 💥**
>
> **Your thoughts aren't facts. They're rough drafts.**

Let's put that into practice. Here are the five biggest brain bugs people run into—and how to debug them fast.

The Big Five Limiting Thoughts — Now Upgraded

"AI is going to take my job."
→ "Parts of my job will change, and I get to evolve right along with them."

"I'm not technical enough."

→ "I'm resourceful—and Google is basically my spirit animal."

"I'm already behind."
→ "Most people are still Googling what AI even stands for."

"This is overwhelming."
→ "One tool, one prompt, one click. I can handle that."

"I don't have time to learn this."
→ "If I learn this now, I get my time back later—compound interest, baby."

Write your own. Make it spicy. Make it stubborn-proof. Stick it where you'll see it daily.

Use Humor to Break the Spell

If you're spiraling, try this:

"I'm not just behind, I'm practically chiseling my emails into stone tablets!"

Or:

"I'm so bad with tech, I once unplugged the router because it was blinking at me."

Exaggeration deflates the fear balloon. You start to see that the voice saying "you can't" isn't a prophet. It's a parrot with a fear complex.

Once you've laughed at the fear, you've taken away its teeth. That's your cue to move—because momentum, not perfection, is the real cure.

When You're Still Stuck, Go for the Micro-Win

You don't need a 10X moment. You need a 1% move.

Open one AI tool. Click a button. Didn't explode? Success. Watch a 60-second tutorial—bonus points if you don't yawn. Ask a question you're "supposed to know"—welcome to being an adult.

Small wins stack. Stack enough and you're not stuck anymore.

The Brainy Bit (Without the Boring)

Every time you reframe a thought, your brain carves a new trail. That's neuroplasticity in action—your built-in upgrade system.

You're not stuck because you're bad at change. You're stuck because your neurons are loyal little soldiers.

But they'll follow new orders if you give them new ones.

Your brain's not broken. It's just waiting for better instructions.

This isn't fluff. Reframing actually changes how your brain fires, which changes how you feel, which changes what you do.

That's how you become the kind of person who says "bring it on" to change, not "get off my lawn."

Fear isn't the enemy. Inaction is.

And that fear you feel? It's just your brain checking whether you're serious about growing or just flirting with the idea.

So let the old thoughts surface. But don't hand them the mic.

Grab your pen. Your voice. Your curiosity.

Remember—rewiring your thoughts isn't about erasing fear. It's about retraining your brain to keep moving even with fear riding shotgun.

And if all else fails, picture me in a Buffalo Bills horn hat whispering:

"The boat's still at the dock. Coffee's hot. You're not late. You're actually early... for the new you."

Now go rewrite something—starting with your next thought.

Chapter 26
Snap Out of the Spiral — Three Mindset Moves That Actually Work

Ever stare at an AI tool and feel your brain quietly power down—like a laptop gasping at 1%? Yeah—welcome to the club.

One day you're cruising; the next, the pressure to figure out AI lands like a weighted blanket stuffed with tech manuals—and guilt. For a second, it's tempting to fake a power-off yourself.

Don't.

I've walked that road—potholes, wrong turns, and the occasional neon sign flashing You're Doing It Wrong!

My goal isn't to pile on more weight; it's to hand you three light, nimble moves I use whenever the tech circus rolls into town. Each one delivers immediate relief.

No deep breaths or yoga mats required.

1. Play Detective with AI — Curiosity Beats Strategy

Play first, plan later.

That may sound backward if you're wired for precision. I used to stare down new tools thinking, I need a battle plan before I touch this. Turns out, the only battle was in my head.

The day I ditched the spreadsheet and just played, everything shifted. I typed, "Write me a love letter to my couch." Ridiculous? Completely. Effective? Shockingly.

That tiny act of nonsense sparked something real—curiosity. The fear melted, and possibility strolled right in. That moment when laughter replaced pressure? That's the pivot.

Funumental Truth Bomb 💥

Play turns the unknown into a playground instead of a minefield.

Your Play Assignment

Spend fifteen minutes in glorious, unproductive nonsense. Ask AI for a pizza topping nobody's tried. Bonus points if it sounds disgusting but delicious. Or have it rewrite your to-do list in pirate dialect.

You're not trying to win—you're trying to un-stick your mind. Low stakes loosen the grip; perfection exits, discovery enters.

Once you realize AI won't explode when you poke it, you're ready for the next move.

2. Call Off the Alarm – Panic Out, Small Experiments In

Most people ignore AI for weeks, then binge twelve videos and try to build a robot army by sundown. That road ends in burnout—and "Learn AI" haunting your to-do list forever.

I quit the blitz—fifteen minutes a day, that's it. No heroics, no crash courses, just one small win.

Timer on. One prompt. Done.

Email summary, blog outline, fake pet-snake name—it doesn't matter. Showing up is the strategy.

Each tiny rep is a brick; alone, small. Together, solid. Before long I went from frozen to fearless—tossing prompts like popcorn, poking buttons like a curious raccoon at a vending machine.

That steady, unhurried progress beats anxiety every time. And once you've got rhythm, you need resonance—people to echo it.

3. Bring Your Crew – Curiosity Grows in Good Company

Here's where most people stall—they try to outsmart isolation.

Learning sticks when you stop pretending you're on a solo quest.

I tried expert forums (all flex, no fun). The real growth came when a few friends said, "Let's learn this together." No pressure, no gurus, just honest experiments.

We shared wins, flops, and the occasional AI-written disaster jingle for an imaginary dating app called Clingr. No judgment—just laughter.

That energy is rocket fuel. Belonging quiets the threat signals that make learning hard. A crew built on curiosity erases the fear that keeps most people stuck.

The Guiding Question

All three moves orbit one deceptively simple question:

"What's one small thing I can do with AI today that makes tomorrow easier?"

Not impressive. Not overwhelming. Just easier.

Maybe it's:

- Practicing one new prompt

- Documenting what tripped you up
- Sharing a weird result with your crew
- Bookmarking a tool (and naming it something ridiculous)

This question cuts through overwhelm like a hot knife through jargon. It trains your brain to favor micro-actions over mythical master plans.

I know because I've built Gantt charts just to avoid clicking a single button. Simple wins. Small is sexy. Done beats dazzling.

While everyone else drafts Prompt Mastery Funnel 7.0, you'll be stacking wins, rewriting your norm, and quietly outpacing them.

Why These Work (The Neuro Nugget)

Play-based learning rewires your brain faster than any manual. Low-stakes experiments lower resistance and boost memory. Small daily actions create sustainable momentum—your brain loves tiny wins and compounds them. Repetition builds confidence without overload, and community cuts fear in half while doubling accountability.

You're not winging it; you're finally working with your brain, not against it.

Try these three moves for a week: Play first. Practice daily. Find your people. Ask what makes tomorrow easier.

You'll stop freezing and start flowing. Momentum will quietly replace overwhelm. And soon you'll be the one who clicks the scary button without blinking—then laughs, looks up, and says,

"Huh. That wasn't so hard."

Now go make one tiny move—then watch what happens.

Chapter 27
Momentum Over Mastery

13 Mindset Moves for People Who Hate "Mindset Work"

Q: How do I stop sizing myself up against everyone who seems way ahead?

A: Comparison is creative kryptonite—social feeds are highlight reels, not behind-the-scenes footage. You don't see the bloopers, breakdowns, or three-hour Canva meltdowns.

The only scoreboard that matters measures your stretch from yesterday. Write down what you couldn't do last month. Track every tiny leap. Celebrate each one like you just won the Entrepreneur Olympics—even if it's just learning one new shortcut.

Momentum favors the self-aware, not the self-comparing. Race your past self, not someone else's algorithm.

Q: What if I feel too old or too stuck to adapt?

A: Oh, I've got the T-shirt for this one. I'm the King of Overthinkers and a card-carrying member of the *Is It Too Late?* Society.

But neuroscience is on your side—your brain still rewires. Neuroplasticity doesn't care about your birth year. You've got pattern recognition the rookies dream of.

You're not outdated software; you're legacy code with the latest patches. You're not stuck; you're just overqualified for fear.

Q: How do I balance healthy skepticism with staying open?

A: Skepticism is a superpower—until it turns into a straitjacket. You don't have to gulp every AI trend like champagne, but if you swat away each new idea like a mosquito, you'll miss the ones that matter.

Ask yourself: is this fact-checking or fear-cosplay?

Stay skeptical, stay curious. Test ideas in small, safe experiments. Think of it as speed-dating for tools: low commitment, high learning, zero wedding invitations.

Q: My workplace treats new ideas like a contagious rash. How do I experiment without getting fired?

A: Classic case of corporate hives. Don't wait for permission—go stealth. Try something small. Share a quiet win. Let curiosity spread like a slow-burn rumor.

Worst case? You get so good they can't afford to lose you.

Best case? You become the change they wish they'd hired.

Q: I keep slipping into perfectionism with every new tool. How do I stop?

A: Perfectionism is procrastination wearing designer shoes. It whispers, "Not yet," until the boat's gone.

I used to whiteboard plans like I was pitching NASA. Now I aim for B-minus work with A-plus consistency.

Funumental Truth Bomb 💥

Progress is messy, glitchy, and wildly effective.

Celebrate your flops—they're feature releases in disguise.

Q: Isn't this "mindset work" just turbo-charged positive thinking?

A: Nope. I don't do unicorn affirmations. Real mindset work is strategy in disguise—figuring out why you're stalling and building a bridge over it.

Think plumbing: unseen, unglamorous, but everything leaks without it.

Identify your mental bottlenecks. Design experiments that make stalling harder than succeeding.

That's not a pep talk—it's performance architecture. Blueprint optional.

Q: I keep thinking I need to master every tool before moving. How do I stop and just start?

A: Mastery is a myth sold by course creators. You don't need mastery—you need movement. "Ready" is a mirage that keeps moving further away.

Cannonball into the sandbox like a toddler with a popsicle. Sloppy action beats pristine theory every time.

Every experiment teaches what prep never will. Done is the new perfect.

Q: My concerns about AI feel legit. Shouldn't I pause until I've analyzed every risk?

A: Respect risk, don't worship it. Fear is a movie trailer—it always looks worse than the film.

Break it down: what's real, what's noise, what's actionable? Then test one small piece.

You're not buying a spaceship; you're dipping a toe. Progress belongs to people who act while others are still analyzing the water temperature.

Q: What if my inner critic keeps yelling that I'm "not enough"?

A: That voice is a glorified heckler. Mine's named Norman. He's a nervous DMV clerk with washed-up boy-band energy.

I thank him for his outdated safety alerts… then do the thing anyway.

Action is how you overwrite that crusty mental code. Let your wins pile up like pop-up ads on a sketchy website until Norman finally shuts it.

Q: Everyone online looks thrilled about AI. How do I tell genuine growth from toxic positivity?

A: Simple. If it sounds like a toothpaste commercial, it's fake. Growth includes groaning.

Real builders share their faceplants, pivots, and chatbot disasters. They talk about the 2 a.m. debugging sessions and the prompts that went hilariously wrong.

If someone's feed is all wins and no wreckage, they're either lying or new. Embrace the mess—that's where the neural rewiring hides.

Q: I keep reading, researching, and scrolling—but my action tally is zero. How do I switch from consuming to doing?

A: Welcome to the Research Riptide. I've drowned there myself. Bookmarks don't build empires.

Every new idea deserves an attached action: read → test → refine. Learning without doing is digital hoarding. You're not building a library; you're building a business.

Set a timer. Fifteen minutes of research equals fifteen minutes of action. One in, one out. Like a cognitive nightclub with a fire code. The bouncer's name is Focus.

Q: I feel awkward asking for help or admitting I'm lost.

A: Join the club. We meet Thursdays and bring snacks. Nobody knows what they're doing; some of us just accessorize better.

Find fellow fumblers. Compare notes. Share flops. Laugh often.

The fastest way forward is together, not alone on Shame Island building sandcastles of regret.

Q: So what's the bottom line?

A: Start smaller than your ego wants. Much smaller. Celebrate "meh" attempts. Reframe doubt as data. Build public flops into your origin story. Find your scrappy squad.

When in doubt—move.

Momentum isn't a muse. It's that stubborn friend who shows up the second you stop scrolling and start clicking.

You're absolutely right - let me show you the refined Chapter 28 directly:

Chapter 28
Key Takeaways From Your Mindset Shift

This isn't a review; it's a checkpoint.

No cheesy credits. No awkward high-fives. No participation trophies.

This chapter connects the dots, celebrates your rewiring, and gets you ready for what's next.

You didn't just read about mindset—you shifted it.

You peeled off labels, tuned out the peanut gallery, and got comfortable with discomfort. Now you're miles ahead of where you started.

Recalibrating How You Handle Change

Let's be honest—panic-scrolling never moved anyone forward. You saw that chasing one more tutorial was procrastination wearing a productivity costume. Trying to "learn everything" while your inner critic yelled "You're behind!" was the express lane to burnout.

You stopped treating mindset like a motivational poster and started using it like a wrench—practical, specific, and occasionally covered in grease.

You practiced curiosity instead of control and caught your brain running on outdated software.

Funumental Truth Bomb 💥

Your thoughts about tech held you back more than the tech itself.

Now you know better. You're running upgraded code. Period.

Your Core Operating System Upgrades

Not fluff. This is your mental tech stack, the psychological operating system that keeps you steady when the apps change.

Emotional Agility: You stopped bulldozing past discomfort and learned to pause, name it, and act anyway. Fear doesn't get veto power anymore—it just gets a seat at the table while you make the call.

Beginner's Mind: You quit waiting to be "ready" and turned learning into a sandbox. Messy? Yes. But now you fail forward on purpose, which beats standing still in style.

Identity Expansion: You peeled your job title off your soul and claimed a new one: Learner-in-Chief. Your worth isn't tied to what you already know—it's tied to what you're willing to try.

Optimistic Realism: You traded perfection for progress. No fantasy deadlines, no guru promises—just steady motion with your eyes open.

Curiosity Over Control: You replaced "I should know this" with "What happens if I try this?" and suddenly work felt playful again. Exploration beats expertise when the map keeps changing.

Experimentation Confidence: You stopped craving certainty and started running experiments. Every test teaches. Every attempt counts. Every flop is data.

None of these are "soft" skills. They're structural. Without them, you loop in anxiety every time tech shifts. With them, you build momentum.

Reframing That Inner Critic

But knowing your new operating system is one thing. Silencing the old error messages? That's where the real work happened.

We didn't delete your doubts—we demoted them.

You turned "I'm not technical" into "I'm learning this, one awkward click at a time." You practiced being fear-wise, not fearless—acknowledge the fear, move anyway.

And how do I know? I'm the King of Overthinkers. If I can do this, you absolutely can.

You remixed your critic's greatest hits until your thoughts sounded less like a doom spiral and more like "Hey, I'm figuring this out." That's the shift.

Action Over Abstraction

You didn't just think differently—you did differently.

You gave yourself fifteen minutes to tinker instead of five hours to spiral. You practiced over perfectioned. You found fellow learners instead of idolizing experts from a distance. You let weird wins count. You let tiny moves matter. You let trying be enough.

That's not soft. That's tactical empathy—for yourself and for the messy process of growth.

Why This Shift Matters More Than Another Tutorial Ever Could

You don't need to be a tech wizard. You just need to stop treating change like a villain.

You've upgraded from fragile to flexible—from "I'll never catch up" to "I've got this moving." The terrain keeps shifting, but now you've got better boots.

You won't always feel ready, but you're now the kind of person who moves anyway. That's the upgrade.

You learned how to navigate fear without letting it captain the ship. You use mindset as a lever, not a label. You make progress in public without waiting for applause or perfection.

You're a mover, a try-er, a builder—someone who experiments instead of excuses, learns faster because you're not afraid to look like a beginner.

What's Next: From Mindset to Motion

This wasn't a pep rally; it was prep school for action.

Next comes the practical stuff—the tools that don't make you want to throw your laptop, the rookie mistakes I've already made so you don't have to, and the frameworks for picking what actually matters instead of what's hyped. You'll make things happen without waiting six months or six certifications.

This is where your brain and the bots shake hands and start building.

You're not a tech genius. You're something better: a momentum-maker who doesn't wait for permission.

Chapter 29
5 AI Tools That Work Harder Than You Do
(No Coding, No Drama)

Ever tried fixing a leaky faucet with the wrong wrench, the wrong tutorial—and a dog actively eating the instructions?

Welcome to how most people feel when they first flirt with AI—it starts cute, then gets complicated.

Cue the confusion montage.

They start out curious, maybe even excited—then they hit a wall of jargon. Tools that look like the control panels of the Millennium Falcon. Suddenly you're Googling "AI for dummies" and questioning all your life choices.

Let me clear this up: You don't need to be a "tech person." That's like saying you need to be a Michelin chef to use a microwave. **Not true, not helpful,** and definitely not how we roll.

Here's what nobody tells you: If you can operate a search bar or set an alarm on your phone, you are more than qualified to dip into the powerful world of AI.

No code. No downloads. No crying into your keyboard.

(And if you ever do want to dabble? There's this thing called "vibe coding" where you basically describe what you want and AI writes the code. But that's for later. Right now, we're keeping it simple.)

Why This Chapter Exists

I've watched brilliant humans—CEOs, coaches, clinic owners, and yes, even a few reluctant uncles—walk away from life-changing tools because the first step felt like a wall instead of a welcome mat.

These people aren't Luddites. They're whip-smart, resourceful, high-functioning people with plenty on their plates already.

And here's what's wild: Every time I've shown one of them the right tool with the right intro, they've bounced back with results like, "That took five minutes?!" followed quickly by, "What else can I automate?"

That's the domino effect—one small win that knocks over a whole mindset.

Momentum is contagious—it starts with a single tiny win. That's what this chapter is about: showing you tools that'll give you your first (or next) "Whoa… that was easy!" moment.

What You're Getting (And What You're Not)

You don't need a museum of AI tools. You need a toolbox. And ideally, one that doesn't require a 17-hour onboarding course and a personal Sherpa.

Each tool has been vetted the Funumental way: fun, fast, and time-giving instead of time-sucking.

If not? Poof—gone.

See ya, Captain Complexity.

No bloated "ultimate lists" of 37 tools you'll never use. No complicated software with 84 tabs and 19 pricing tiers. No pretending shiny equals useful.

A Quick Word from Your Favorite Reformed Overthinker

Look, I get it. I used to be the King of Overthinkers. My inner skeptic could win gold medals in mental gymnastics and probably a silver in procrastination.

I've stared at dashboards like they were cryptic escape room puzzles. But once I stopped trying to master everything before I clicked anything… everything shifted.

So let's make a deal. Don't wait for perfect clarity.

> **Funumental Truth Bomb 💥**
>
> **You don't need mastery to start. *You need one click.***

Tiny wins equal big confidence. You don't need to be brilliant to begin. You begin, then you become brilliant.

What You'll Actually Learn

You'll learn how to spot tools that actually solve your problems—not just look cool on YouTube. You'll see where each tool shines, where it flops, and when to walk away. And you'll get a single first step to test each one, because thinking about it is not the same as trying it.

Bonus: You'll learn to evaluate tools on your own moving forward. Consider this the AI equivalent of teaching you how to order off-menu at a restaurant—without needing to learn the whole cookbook.

Order confidently. Leave the chef hat to the algorithms (and tip your data).

What This Really Unlocks

Let's be clear: this isn't about becoming an AI guru or learning Python between Zoom calls.

This is about finally reclaiming space—in your calendar, your brain, and your nervous system—for the stuff that actually matters. Strategic thinking. Creative flow. Rest that doesn't feel like a productivity crime.

This is where you stop being the duct tape holding your business together with caffeine and hope—and start showing up as the visionary: the architect, the actual adult in the room with a game plan instead of a Google Doc full of chaos.

These aren't novelty tools. Think of them as your delegation team—minus the HR paperwork. They not only make your life easier—they make you feel like you again. The version of you who gets things done without needing to sacrifice your sanity, your sleep, or your last good nerve.

Your First Tool (No Prep Required)

Next up: your first AI tool. Not one of those "watch this 47-minute tutorial before you breathe" tools. We're talking: "Damn… this just gave me back an hour of my life."

No pressure. No prep work. Just real-world, usable, high-five-your-future-self kind of magic.

And if your inner critic starts whispering, "I don't know enough…"—smile and keep moving. That voice is your old operating system trying to reboot. Let it buffer. You've got better tools waiting—and work worth doing.

Chapter 30
Your AI Starter Pack
(aka: The Tools That Won't Ghost You After One Use)

So let's do the thing.

Stepping into AI can feel like walking into Costco—every aisle promising to change your life, when you only came in for batteries.

I've watched brilliant humans freeze like a browser with 47 tabs open—all because they had no idea where to begin.

So I made you a map. A beginner's compass. A golden short list of tools that don't just crack the door open—they roll out the welcome mat, hand you a latte, and say—

"Let's do this."

Why Fewer Tools = More Results

Every time someone shares their AI stack online, I brace myself for the inevitable mega-list that ends with "oh and don't forget these hidden gems."

Hard pass.

That's not a strategy. That's a scavenger hunt for your next burnout.

Here's what actually helps: a tight toolkit. Five AI tools. That's it. Period.

And if you really want to ditch overwhelm, just start with the first one.

Funumental Truth Bomb 💥

Fewer tools. More results. Period.

Each of these tools solves a different kind of problem. Each one delivers real value. And the best part? No coding. No chaos. No crying in the settings menu.

Tool #1: ChatGPT

Your Brainstorming Beast Buddy

Let's start with the one everyone's heard of.

I once hoped ChatGPT would ghostwrite my next book while I binge-watched vintage Star Trek. That fantasy? Vaporized by day two.

But ChatGPT did become my go-to for breaking through mental grid-lock. Want to punch up a headline? Draft a less-awkward email? Summarize a podcast like you actually took notes on it? ChatGPT does all that with zero attitude and no caffeine dependency.

Pro tip: Talk to it like a nerdy friend who always has ideas. You'll sound sharper, faster, and funnier in five minutes flat.

First win: Ask it to rewrite a boring bio. You'll grin. You'll tweak. You'll wonder how you ever worked without it.

Tool #2: Claude

The Sensitive Genius of the AI World

Next up: the thoughtful one.

Claude is that friend who brings snacks and emotional support on your worst day. When tone matters—like really matters—Claude gets it.

Writing something heartfelt? Ethical? Slightly soul-baring? Claude won't strip the soul out of it. It keeps the human in your humanity.

Use it when: You're writing a client apology, a vulnerable story, or anything where the how matters as much as the what.

When empathy matters, Claude delivers—no tissues required.

ChatGPT is your brainstorm buddy. But Claude? Claude is your writing therapist.

Tool #3: Perplexity

Instant Genius Without the PhD

Now for the fact-checker.

Ever wanted Google to just give you the answer—with sources? That's Perplexity. It's your fact-checking friend who doesn't make stuff up and always shows their work.

Ask it anything: stats, summaries, explanations. Perplexity will spit back crisp answers plus clickable sources. It's like hiring a research assistant who works for praise alone and doesn't need coffee breaks.

Start with: Something you've been curious about but too busy to Google. Watch it deliver clarity in seconds. Then copy one answer into ChatGPT and ask it to simplify it further—you'll see teamwork in action.

Tool #4: Canva Magic Studio

Design Wizardry Minus the Wand

Time for the creative tool that doesn't require creativity.

Before Canva, I thought graphic design required a secret handshake and a MacBook that cost more than my car. Now? I can make slick graphics in yoga pants with coffee stains and confidence.

Canva Magic Studio gives you the power of a full design team—minus the group Slack channel. Slides. Reels. Social posts. Flyers for your left-handed knitting club. You name it, it makes it pop.

Pro tip: Use it to design one thing—just one—and you'll feel the shift. Canva confidence is a real thing.

Tool #5: Notion AI

Your Brain, But With a Search Function

And finally, your digital brain assistant.

Notion AI is like hiring a project manager for your brain. You toss in scattered notes, rogue ideas, "remind me to follow up" blurbs—and it turns them into organized magic.

Summarize. Tag. Prioritize. Build from the chaos.

Use it for: Messy meeting notes, voice memo transcripts, or the back-of-napkin brilliance you usually forget by dinner.

You don't have to be "naturally organized" anymore. Now you just have to show up and toss it in.

It doesn't replace your brain; it organizes it so you can actually use it.

Making It Stick

Now that you've met your crew, let's talk about actually using them.

Because knowing about tools and using tools are two very different things.

Clicking around is fun. But clicking around doesn't get stuff done.

Pick one real task. Use one real tool. Get one real win.

Need a caption? Ask ChatGPT. Cluttered notes? Dump them in Notion AI. Just start somewhere. You don't need to learn the tool. Let the tool help you solve something real.

I've seen beginners double their productivity using just two of these tools. Not because they memorized tutorials, but because they dared to use them on real problems.

Small wins stack fast. A better email. A cleaner slide deck. A fact-checked stat. You stop resisting and start exploring. Momentum's contagious—and now you've got proof.

This isn't a scavenger hunt anymore. It's a shortcut—with better snacks and fewer breakdowns.

Pick one tool today. Fix one thing that annoys you. Celebrate your tiny, mighty win—then brag about it to your future self.

You've got your AI Starter Pack. Now go use it—and watch what shifts.

Chapter 31

Decision Guide: Where to Start

(aka: Finally Get Sh*t Done)

You've got your starter pack. Now comes the hard part—actually using it.

This is where we trade the headlights for high beams and finally pick a lane.

This chapter isn't about mastering every feature or memorizing shortcuts like you're prepping for *Jeopardy—Tech Bro Edition.* It's about figuring out which tool to use right now—without melting your brain in the process.

We're ditching the digital treasure map and handing you the keys to the fastest ride out of indecision.

The Trap of Tool FOMO (and Why More Isn't Better)

Let's be real: every week, someone posts a thread titled "My 47 AI Tools That Changed Everything" and suddenly your poor to-do list is weeping in the corner while you're opening tab 26 of 94.

That's not productivity. That's tab chaos masquerading as research.

Funumental Truth Bomb 💥

You don't need more options. You need traction.

Golden Rule: Don't pick tools based on hype. Pick them based on the actual task sitting in front of you.

Decision GPS: Start With What Sucks Most

Skip the fantasy planning. Skip the pretty dashboards. Ask yourself: What's the one annoying thing I'm avoiding right now?

Because your brain rewards motion, not speculation.

Messy ideas? Use brainstorming tools.
Finding facts? Use research tools.
Turning chaos into clarity? Use organization tools.

That's your starting point.

You're not marrying these tools. You're speed-dating them for the job at hand.

The Four Doors: Your Cheat Sheet to Momentum

Think of it like this: every task you face walks through one of four doors. Pick the right door, and you're golden.

Door #1 – Brainstorm / Draft

Tools: ChatGPT or Claude

They don't judge your half-baked thoughts—they just roll with it. Use ChatGPT for fast first drafts and rapid-fire ideas. Use Claude when tone matters and empathy needs to sneak in between the words.

Door #2 – Research

Tool: Perplexity

Gives you facts with sources, not hallucinations with confidence. No BS—just answers with clickable sources so you can finally close half those tabs.

Door #3 – Design

Tool: Canva Magic Studio

Makes non-designers look like they've been secretly moonlighting at Adobe. Slick design, low effort, no design degree needed.

Door #4 – Organize

Tool: Notion AI

Turns your mental clutter into task lists with receipts. Chaos becomes clarity, and your ideas finally stop dying in the notes graveyard.

These aren't theory toys; they're practical levers.

Each one solves a real-world pain point—not a theoretical puzzle.

Pick a door. Walk through it. Worst case? You learn. Best case? You fly.

The Cure for Perfection Paralysis

Reality check: you're not here to impress your high school computer teacher. You're here to solve real stuff, and guess what? Every tool has an undo button. That means no excuses.

Click the wrong thing? Great—you just learned what not to do. You're not failing—you're calibrating.

Confidence doesn't come from knowing everything; it comes from doing one thing—and realizing you didn't die.

Treat your tools like a playground, not like a final exam. Press buttons. Laugh at the weird results. Tweak, repeat, win.

That's not trial and error—it's trial and evolution.

Real World Flow: Watch This Magic

Let's see it in motion.

Say you need to write a pitch. Here's how to stack the tools like a Funumental pro:

Start in ChatGPT to draft the bones. Grab stats from Perplexity so you don't sound like you're making it up. Add visuals in Canva Magic Studio so your slides don't scream "2006." Use Notion AI to turn feedback into action items instead of forgotten Post-its.

One of my clients used this exact flow to prep for a board pitch in 48 hours—she landed the deal, slept for twelve hours, and woke up a believer.

That's not hustle—that's harmony.

That's not a tech stack. That's a flow state.

Your Personal Cheat Sheet

(Tape this above your monitor)

Make it simple. Make it yours. Stick it somewhere visible:

- Idea explosion → **ChatGPT** or **Claude**
- Need stats → **Perplexity**
- Ugly slides → **Canva**
- Messy notes → **Notion**

Just don't use 12 flying Post-its like I did. That road leads straight to chaos (and very yellow).

Stop Waiting for the Perfect System

You're not building the perfect system; you're cleaning up your digital kitchen so you can actually cook something today.

Try a tool. Toss it if it sucks. Keep the one that clicks. Your needs will evolve. So should your setup.

This isn't about becoming an AI master. It's about firing your inner procrastination goblin and finally getting something DONE today—without crying into your Google Drive.

Every time you use the right tool for the job, you make your future self's day a little easier. A little clearer. A little more powerful.

You don't need a PhD in prompt engineering. You just need one moment of clarity and the guts to press Go.

Momentum doesn't need your perfection—it just needs your participation.

Pick your door. Walk through it. Get weird, get working, and watch what you build.

Chapter 32

Make AI Give a Damn (Because You Do)

You've picked your door.

Now let's make sure what's behind it actually matters.

I once used AI to survive a conversation I'd been dreading for weeks. Not a "which-tool-should-I-use" question — a "how-do-I-say-this-without-imploding-my-business-partnership" crisis.

Every draft landed too harsh or too hollow. My stomach turned every time I opened the email app.

Finally, I dumped it all into Claude:

> "Help me tell my business partner their communication style is making me want to hide under my desk — but I still think they're brilliant and want to fix this."

The AI didn't solve it for me — but it helped me *feel* my way to clarity.

The talk went better than I dared hope. We're still partners, and I no longer wake up dreading my inbox.

That's when AI matters — when the outcome hits you in the feels.

Why Emotional Stakes Matter

This isn't about the tool.

It's about what the tool makes possible.

You don't need another shiny app. You need a *win.*

A real-world, *"holy sh*t, this just bought me my sanity back"* kind of win — not another digital paperweight collecting dust while you second-guess your life choices.

Funumental Truth Bomb 💥

Emotional stakes always win. Logic is for spreadsheets. Feelings are for humans.

Most folks fall into the *"ooh cool demo"* trap. They smile, maybe chuckle at the robot's attempts at humor — and then never open it again. Why? Because it didn't matter to them.

If it doesn't solve something *personal* and *annoying,* I don't care if it's been trained on every book ever written. You're not here for cat poems. You're here to stop dreading email like it's jury duty.

When fun meets function, friction fades.

That's when AI finally starts giving a damn.

Trying to learn every tool upfront is like buying a chainsaw to cut a bagel.

It's loud, scary, overkill — and your bagel looks like it lost a fight with a weed whacker.

You don't need to master everything before breakfast. You need to solve something.

That nagging decision you've been avoiding. That text you keep rewriting like it's the final scene of a hostage negotiation. That spreadsheet you've opened thirty-seven times and still haven't updated because it drains your soul faster than a meeting that should've been an email.

Pick one. Feed it to AI.

Then don't stop until you get relief.

That's how you get hooked — not on the tech, but on the *feeling* of momentum returning.

Make AI Personal

When AI helps you win small, you start to trust it.

And when you trust it, you start experimenting for real.

Use it where the stakes feel emotional — not abstract.

- Draft the email you've been avoiding.
- Polish the offer that still doesn't sound like you.
- Write the apology you owe yourself or someone else.
- Rehearse the tough conversation with the client who ghosted you.

That's when it clicks: AI isn't replacing your humanity — it's reflecting it back to you, cleaner and faster.

The goal isn't to sound like a robot that learned manners.

It's to use the robot so *you* can sound more human.

Feel First, Format Later

People waste hours trying to make AI sound professional before it sounds *real.*

They'll tweak syntax, swap adjectives, and polish commas while missing the actual emotional beat.

Here's the reframe: **Start with the emotion. Format it later.**

Tell AI what you're feeling before you tell it what you want.

Say things like:

> "I'm anxious about sending this."
>
> "I want this post to sound confident but not arrogant."
>
> "Help me explain this without sounding defensive."

You'll be amazed how much better the outputs get once you bring the *why* before the *what.*

Because emotion is context — and context is where the intelligence lives.

> **Funumental Truth Bomb 💥**
>
> **AI mirrors what you feed it. Garbage in, garbage out — but honesty in, brilliance out.**

When you let it see your tension, your humor, your humanity — it responds in kind.

That's when it starts *giving a damn* because *you* did first.

So stop hiding behind the perfect prompt and start showing up like the messy, magnificent human you are.

Type like you talk.

Ask like you care.

Experiment like nobody's watching.

AI doesn't need you to sound smart.

It needs you to sound *real.*

The Real Win

AI's greatest gift isn't speed.

It's reflection.

It shows you what you really think when you stop censoring yourself. It's the mirror that answers back — the one that helps you write your way to truth.

So yes, make AI give a damn.

Because when it does, what it really mirrors back… is that *you still do.*

Chapter 33
AI as Mirror (and How to Actually Use It)

You don't need another tool.

You need a mirror that answers back.

That's what AI can be when you stop treating it like a vending machine for cleverness and start using it like a coach, a thinking partner, and the most patient interviewer you've ever met.

Most people ask AI for outputs.

Smart people use AI for insight.

And here's the secret: insight beats output every time.

Why a Mirror Beats a Megaphone

You already have opinions, blind spots, patterns, and strengths you can't see because—you know—you're inside your own head. AI won't hand you your identity on a silver platter. But it *will* reflect what's already there, faster and clearer, if you ask it the right questions.

> **Funumental Truth Bomb** 💥
>
> **If you only ask AI for answers, you'll miss its questions.**

The right mirror doesn't flatter.

It clarifies.

Four Mirrors You Actually Need

1) The Clarity Mirror (Get What You Mean… to Say)

Use when: you're fuzzy, rambling, or overexplaining.

Prompt:

> "I'm trying to say this, but I'm not landing it.
>
> Here's my messy draft: [PASTE].
>
> Summarize my core point in one sentence, then give me three sharper versions with different tones: (a) warm/confident, (b) direct/brief, (c) playful/clever.
>
> Tell me what I *think* I'm saying vs. what I'm *actually* saying."

What you'll see in the mirror: the gap between intention and impact.

2) The Value Mirror (Does This Sound Like Me?)

Use when: your copy is technically great but feels… not you.

Prompt:

> "Here are five things I value in my voice: [LIST—e.g., human, witty, clear, kind, no fluff].
>
> Here's a sample that *does* sound like me: [PASTE].
>
> Now revise this new piece to match those values: [PASTE].
>
> Explain what you changed and why, in bullet points."

What you'll see: your voice codified—so you can reuse it.

3) The "Yeah-But" Mirror (Challenge My Thinking)

Use when: you're confident… maybe too confident.

Prompt:

> "Play devil's advocate on the argument below.
>
> Identify weak assumptions, missing data, and where bias might be creeping in.
>
> Then steelman my position—make the strongest possible version of the argument I'm trying to make.
>
> Finally, give me the two best counterarguments I should be ready for.
>
> Text: [PASTE]."

What you'll see: your blind spots—without the ego bruise.

4) The Future-You Mirror (Run the Rehearsal)

Use when: you've got a high-stakes conversation, pitch, or boundary to set.

Prompt:

> "Role-play as [WHO: my client / my boss / my business partner].
>
> Use their likely tone and priorities.
>
> I'll write my side; you respond as them.
>
> After three exchanges, switch to coach mode and score my clarity (0–10), empathy (0–10), and credibility (0–10).

Suggest one line that would have landed better."

What you'll see: the reps you needed *before* the moment.

The Funumental Mirror Workflow (10 Minutes, Real Results)

1. Dump (2 min): Paste the messy version. No polishing.
2. Distill (2 min): Ask for a 1-sentence essence + 3 tone options.
3. Diagnose (2 min): Ask, "What's unclear, off-brand, or likely to trigger objections?"
4. Dialogue (2 min): Role-play pushback; iterate lines in real time.
5. Decide (2 min): Choose one version. Ask for a final polish in your voice.

Funumental Truth Bomb 💥

Don't chase perfect. Chase *true*, then make it clear.

Prompts You'll Actually Use This Week

- Mirror my message: "Here's my draft. Extract the emotional core in one line, then rewrite for (a) a skeptical reader, (b) a busy reader, (c) a friendly insider."
- Turn my ramble into slides: "Turn this into a 7-slide outline with titles, 3 bullets per slide, and one story beat per slide: [PASTE]."
- Make it sound like me (not a press release): "Rewrite this to sound human, not corporate. Shorter sentences, less fluff, same meaning. Keep my humor if present. [PASTE]."
- Ask me better questions: "Based on this situation, ask me 10 precise questions a great coach would ask to clarify my goal, constraint, risk, and next step. [PASTE]."

- Ship it now: "Give me a version I can send as-is in 120 words or less. Start with the point, no throat-clearing. [PASTE]."

Common Mirror Mistakes (And Quick Fixes)

- Mistake: "Make this better."

Fix: Specify *how*—tone, audience, constraint, desired feeling.

- Mistake: Writing for approval, not impact.

Fix: Lead with the *why* ("I'm nervous," "This is time-sensitive," "I want calm + clarity.")

- Mistake: One-and-done prompting.

Fix: Iterate. "Try again, but…" is where the gold hides.

- Mistake: Asking for "professional" (and getting sterile).

Fix: Ask for *clear, kind, concise*—then add your values list.

Make It Matter (Tie to a Real Stake)

Pick one emotionally sticky task today:

- The overdue email that's now a brick in your stomach
- The pricing note you keep punting
- The "we need to talk" message
- The apology / boundary / renegotiation

Run it through the Funumental Mirror Workflow.
Ship the final.

Then notice: it wasn't the tool that changed you.
It was the clarity you got while using it.

The Line Between You and the Robot

AI can predict the next likely word.
You can choose the next *true* one.

AI can mirror your voice.
Only you can own your values.

AI can draft your message.
Only you can stand behind it.

Funumental Truth Bomb 💥
AI scales whatever you bring to it. Bring courage.

Because when the mirror shows you something real, your job isn't to admire it.

It's to act on it.

Your 48-Hour Challenge

1. Pick one high-friction task.
2. Run the 10-minute workflow.
3. Send it.
4. Journal one line: *"What did the mirror show me about me?"*

Do that twice and you'll never look at AI as "just a tool" again.

You'll see it as the mirror that helps you mean it.

And once you mean it, momentum shows up.

Keep AI talking human to you—and let the mirror make you braver.

Chapter 34
Your Next 30 Days
(From Learning Mode to Living It)

You've dodged the landmines. Now let's build momentum that sticks—whether you're flying solo or leading your crew.

Momentum isn't built by magic. It's built by motion, repetition, and the courage to keep showing up.

Here's the uncomfortable truth about learning anything new: the first week feels like magic. Week two? Still exciting. By week three, the shine wears off—and that AI tool sits in your bookmarks like an expensive gym membership.

This chapter isn't about inspiration. It's about integration. How do you take what you've learned and actually make it part of how you work, think, and solve problems? How do you help your team do the same?

> **Funumental Truth Bomb** 💥
>
> **You don't build habits through willpower. You build them through design.**

The Shift Nobody Talks About

There's a moment that separates people who "tried AI once" from people who can't imagine working without it. It's not when you get your first perfect output. It's when you stop thinking about the tool and start thinking about the outcome.

You don't think about your keyboard when you type. Or your coffee maker when you need caffeine. That's where we're headed—AI so natural it fades into the background—like your keyboard or coffee maker.

The question isn't "Should I use AI for this?" It becomes "Which part of this do I actually need to do myself?"

For leaders, the question becomes: "Which parts of my team's work should stay human, and which parts can AI accelerate?"

Your First 30 Days: A Realistic Roadmap

Forget the "master AI in 21 days" nonsense. Here's what actually works, whether you're implementing this for yourself or rolling it out across your organization.

Week 1: Anchor to One Repeatable Win

For individuals: Pick ONE tool. Use it for ONE specific thing. Every single day.

Repetition beats brilliance when you're rewiring behavior.

Not "I'll try different things and see what sticks." That's tool tourism—fun for Instagram, useless for results. You want one reliable win that becomes muscle memory.

For teams: Pick one pilot group (3-5 people) and one clear use case. Let them test for the whole week. Document what works.

Examples: Use ChatGPT to draft every email reply before you send it. Run every creative brief through Claude before your team sees it. Check Perplexity before you make any claim in a presentation. Design every social post in Canva instead of overthinking it.

The goal: By day seven, using the tool should feel automatic, not aspirational.

Week 2: Add the Second Layer

For individuals: You've got one habit locked in. Now add a second tool for a different need.

For teams: Expand your pilot group OR add a second use case for the same group. Don't go wide yet—go deep.

The key: Don't abandon tool #1. Stack the new habit on top.

Maybe you're using ChatGPT for emails. Now add Notion AI to organize the ideas that come out of those conversations. Or use Perplexity to fact-check the content ChatGPT helped you draft.

Watch for the multiplier effect. Tools start working together. Your workflow gets smoother. You stop thinking "I should try AI" and start thinking "I couldn't do this without it."

Week 3: Teach Someone Else

For individuals: This is where it gets real. Can you explain to another human how you're using these tools and why it matters?

For teams: Your pilot group becomes your teachers. Have them share what they've learned—no slides, no scripts, just real talk. Show real before-and-after examples.

> **Funumental Truth Bomb** 💥
>
> **If you can't explain it, you don't own it yet.**

Show a coworker. Walk a friend through your process. Record a voice memo to yourself explaining your workflow. Teaching forces clarity. Clarity breeds confidence.

Plus, when you help someone else get a win, you'll remember why this matters in the first place.

Week 4: Optimize One Thing

For individuals: By now, you're using AI regularly. But are you using it *well*?

For teams: Collect feedback. What's working? What's frustrating? What needs clearer guidelines?

Pick one area where results are inconsistent and dig deeper. Refine your prompts—add more context and specificity. Try a different tool for the same task. Automate something you've been doing manually. Build a template or saved-prompt library for recurring tasks.

This is where good becomes great. You're not just using AI—you're turning friction into flow.

How to Know You're Making Progress

Forget tracking hours saved or tasks completed. Here are the real signs you're building momentum:

You stop asking "Can AI do this?" and start asking "How should I phrase this?"

You feel weird when you can't use it. Like leaving your phone at home. It's not a novelty anymore—it's part of your toolkit.

You start seeing opportunities everywhere. That annoying weekly report? AI could draft it. That brainstorm meeting that goes nowhere? AI could help structure it.

You're helping others. When people ask "How'd you do that so fast?" and your answer includes AI, you're there.

For leaders: Your team shares tips spontaneously. People ask "What tool did you use for that?" instead of "How long did that take you?"

When You Hit a Wall (And You Will)

First, breathe. Everyone hits walls. I've rage-quit more AI tools than I care to admit.

Momentum isn't linear. You'll have days where AI gives you garbage. Weeks where you forget to use it. Moments where you wonder if this is worth it.

That's not failure. That's feedback.

Here's what to do:

When results disappoint: Check your prompt. Are you being specific enough? Does the AI have enough context? Would a human understand what you're asking?

When you forget to use it: Tie it to something you already do. Check email? Prompt ChatGPT first. Start a project? Open Notion AI. Make it impossible to skip.

When you feel overwhelmed: Go back to ONE tool for ONE task. Simplify. Reset. Rebuild from there.

When you wonder if it's worth it: Calculate what you got back—time, sanity, creative energy, team capacity. It's probably more than you think.

For leaders: If adoption is slow, don't mandate it. Find your champions, amplify their wins, and let curiosity do the rest.

The Long Game: Building a Sustainable Practice

Here's what separates the dabblers from the Integrationists—people who integrate AI into their actual lives and businesses:

They don't chase every new tool. They master what works and ignore the hype.

They build systems that protect habits. Habits require willpower. Systems just run.

They stay curious, not compulsive. They'll try new features when they matter, but they're not refreshing Product Hunt every morning.

They know when NOT to use AI. Some things need your human brain, your intuition, your lived experience. AI is powerful, but it's not wisdom.

For leaders: They create environments where experimentation is safe, documentation is valued, and curiosity is rewarded.

> **Funumental Truth Bomb** 💥
>
> AI won't replace you. But someone using it **meaningfully** will move faster—and you can be that someone.

Your Commitment (To Yourself or Your Team)

By the end of these 30 days, you won't be an AI expert. You'll be something better: an AI user who gets results.

You'll have tools you trust. Workflows that work. Wins you can point to. And if you're leading a team, you'll have champions who can help others get there too.

When someone asks "Is AI actually useful?" you won't need to theorize. You'll just show them.

You don't need another manual—you need momentum.

Start tomorrow. Pick one tool. One task. Thirty days. That's it.

Not to become perfect. Just to become the person (or team) who stopped wondering and started doing.

The next 30 days are yours. Stop planning. Start using. Watch what shifts.

Chapter 35
The Real-World Leap

You've mapped your first 30 days. Now let's talk about what comes next.

You made it—and that deserves a standing ovation. Not the slow clap kind—the full "we survived another software update" kind. You didn't just skim headlines or let AI hype wash over you like background noise. You engaged. You questioned. You experimented. You let the weirdness happen.

That's where the learning really started.

Somewhere in the glorious mess of curiosity, chaos, and clicking "Regenerate" way too many times, something shifted.

That shift? It's not theory anymore—it's you in motion.

You're not just thinking about how AI might change things—you're feeling it. You're the one nudging levers, testing boundaries, and seeing what's possible. That's the difference between a passive reader and an active creator. You didn't just study the map. You started walking the terrain.

And let's be honest—that's a big freakin' deal.

The Invisible Wall Between Knowing and Doing

You know that weird invisible barrier between "Yeah, I get this" and "Wait—why am I not doing it?"

It's the gap between insight and impact. Between "I get it" and "I'm doing it."

That's the wall. The one between understanding and action.

It's not made of concrete or code. It's built from second-guessing, option overload, and way too many browser tabs.

It shows up when you're excited but frozen, thinking:

Which tool do I actually use?

Am I doing this right?

What if I look like an idiot while I figure this out?

This wall doesn't mean you're not smart, creative, or capable. It means you're human.

We don't get stuck because we're lazy; we get stuck because we care—because we're afraid of looking like fools while we figure it out.

> **Funumental Truth Bomb** 💥
>
> *Caring means you're invested. Just don't let caring become caution.*

But that hesitation? That's where progress tries to sneak past you in the night.

> **Funumental Truth Bomb** 💥
>
> *You can't think your way through the wall. You have to move through it.*

So what do we do?

We don't bulldoze through the wall. We get curious. We tinker. We take one tiny, delightful, "let's see what happens" step forward.

That's the secret.

Momentum loves movement. Let's keep it going.

So Here's the Deal

I could end the book here and wish you luck. But we both know what happens next—you'll feel inspired for 48 hours, then life gets loud, and that bookmark stays right where you left it.

I don't want that for you.

That's why I created two ways to turn everything you just learned into **one working AI system that actually saves you time, cuts costs, or drives revenue.**

Not theory. Not someday. Not a Notion board full of good intentions.

A real system. In your real business. Producing real results.

Welcome to **The AI Integrationist** experience—where the rubber meets the road, and you become the person who doesn't just talk about AI, but actually uses it to win.

Choose Your Path Forward

AI Profit Blueprint (Self-Paced)

For the self-starters who love structure but want to move at their own pace.

This is your 30-day system to go from "I know AI could help" to "Here's the AI system I built that's already working."

Here's What You Get:

✓ **30 Daily Implementation Prompts** - Follow the exact roadmap from Chapter 34, one step at a time

✓ **Dr. V's Prompt Vault** - 50+ battle-tested prompts for emails, content, research, and strategy

✓ **Tool Selection Matrix** - Know exactly which AI tool to use, when, and why

✓ **Video Walkthroughs** - Watch me use these tools in real business scenarios

✓ **Ethical AI Guidelines** - Stay human while scaling with AI

✓ **Progress Tracker & Accountability System** - See your momentum build week by week

✓ **The AI Integrationist Community** - Get unstuck fast with other book readers on the same journey

Perfect if: You're self-driven, love having resources at your fingertips, and want to knock this out on your own timeline without waiting for a cohort to start.

AI Profit Sprint (Live Cohort)

For the ones who thrive with accountability, live coaching, and a crew who gets it.

This is everything in the Blueprint—plus the human connection, real-time troubleshooting, and "let's build this together" energy that makes implementation actually stick.

Everything in Blueprint, PLUS:

✓ **Live Weekly Sessions with Dr. V** - Get your questions answered in real time

✓ **Hot-Seat Coaching** - Bring your actual business challenges and we'll solve them together

✓ **Accountability Partnerships** - Pair up with fellow Integrationists who keep you moving

✓ **Real-Time Feedback** - Submit your prompts, workflows, and outputs—get expert eyes on them

✓ **Priority Access** - First dibs on new tools, templates, and bonus trainings as they drop

Limited to 25 participants per cohort so everyone gets real attention, not a webinar experience.

Perfect if: You want guidance, camaraderie, and someone to look you in the eye (virtually) and say, "You've got this—now let's do it."

Founding Members

If you're reading this soon after the book launches, you may qualify for **Founding Integrationist rates and lifetime benefits.** These are one-time offers that won't be available again once the founding cohort fills.

Check the landing page for current availability—spots fill fast, and when they're gone, they're gone.

My Personal Guarantee to You

If you don't have **at least one AI system integrated into your daily workflow**—if you're not seeing real time savings and better outputs—I will work with you until you do.

Because this isn't another passive course—it's personal.

But you have to put the work in. Take the reps. Show up. That's it.

This isn't magic—it's momentum. I'll give you everything you need and support you every step of the way, but you've got to move.

That's my promise. You don't fail. We don't quit. We keep going until you get results.

Ready to Become an AI Integrationist?

Start here: funumental.com/talk-human-to-me

Use code **TALKHUMAN** at checkout - Your exclusive book reader bonus unlocks special resources and perks available only to readers.

P.S. The landing page will help you choose between Blueprint (self-paced) and Sprint (live cohort) based on your learning style, budget, and timeline. Most people start with Blueprint and can upgrade to a future Sprint anytime—but if you want the founding member benefits, don't wait.

P.P.S. This is your first step as an AI Integrationist. Once you've built your first working system, the Funumental ecosystem opens up—programs for business growth, leadership, and transformation. And when you're ready to scale with total clarity, Crucial OS is there for business owners who want to build profitably while reclaiming their time. Start here. Build from here.

You Ready for This?

You didn't wait for someone to grant you permission. You granted it to yourself.

And now you're more than surviving the AI revolution—you're thriving in it. With curiosity, courage, and maybe a few snarky prompt rewrites that made you giggle.

This is your real-world leap.

You've got the tools. You've got the framework. You've got the mindset.

Now you just need to keep moving.

Not perfectly. Not flawlessly. Just consistently.

Funumental Truth Bomb 💥

One prompt at a time. One win at a time. One day at a time.

The bots are ready. Are you?

Your next chapter is already waiting—now go raise a few eyebrows, rewrite a few prompts, and prove that the smartest thing about AI... is still you.

Keep AI talking human to you—let's see how far you'll take it.

Acknowledgments & Appreciation

Completing *Talk Human to Me* has been a journey of discovery, inspiration, and the occasional need for caffeine and a nap. I am grateful to everyone who walked alongside me, offering support, wisdom, and encouragement.

To my wife, Maggie Unzueta: You're always the real MVP of all my projects. Your love, patience, and perfectly-timed "Go to bed" reminders kept me sane. Without you, this book would still be a pile of half-baked ideas and four walls of rogue sticky notes and half-finished metaphors. Thank you for believing in me and always knowing when to hand me a coffee—or, more likely, gently pry it out of my hands.

To my stepson, Jack: Your boundless creativity and dedication to your music remind me daily to dream bigger and approach everything with curiosity and fun. You remind me to stay curious—and never take myself too seriously.

To Rich Schefren, Neal Schaffer, Molly Mahoney, Nika Stewart, Anik Singal, Jon Benson, and John Paduchak: Your guidance has been the GPS for my wandering ideas. Thank you for your expertise, your patience, and for tolerating my bad puns and stepdad jokes.

To my friends, colleagues, and the entire Funumental community: You've been my cheerleaders, my sounding boards, and my motivation. Your enthusiasm and belief in this mission have fueled me more than any triple-shot espresso ever could.

And to you, the reader: You're the reason this book exists. Your drive to create freedom, lead with creativity, and use AI as your mirror, copilot, and amplifier inspired every page. I hope this book helps you turn dreams into action, spreadsheets into smiles, and adopt TGIM (Thank God, It's Monday) as your motto.

Thank you all for proving that the best journeys are never taken alone—and that even the most chaotic drafts can become magical with the right people by your side.

With gratitude and a wink,
Dr. Christopher S. Vogelmann, D.C. (aka "Dr. V.")
Coronado Island, San Diego County, California

Epilogue: What Comes After Integration

You've done something most people won't.

You read the book. You questioned your assumptions. You let yourself imagine working differently. You didn't just consume information—you engaged with possibility.

That's not nothing. That's everything.

But here's what I've learned after years of helping people transform how they work, think, and lead: insight without integration is just entertainment.

You can know all the right things and still run the same old operating system. You can have the tools and still default to the patterns that keep you stuck. You can want change and still choose comfort when the moment arrives.

The AI Integrationist Program gets you moving. It builds habits. It creates wins. It proves you can do this.

But real transformation? That happens when you stop using new tools and start becoming someone new.

Funumental Truth Bomb 💥

Revolutions don't come often. The AI revolution is the spark. Your transformation is the integration.

Think of it this way: You've been upgrading your apps. Now it's time to upgrade your operating system.

It's where habit becomes identity. Where skills become instinct. Where the person you're becoming finally matches the vision you've been carrying.

I built the AI Integrationist Program for people who finish this book and realize they're just getting started. For the ones who taste momentum and want to sustain it. For the leaders, creators, and entrepreneurs who know that to change their business, they first have to change how they work.

From there, the Funumental ecosystem opens up—programs for business, health, wealth, and transformation. Some will join community programs for ongoing momentum. Others will explore advanced workshops. And when you're ready for the deepest work—the kind that rewires identity, not just workflow—Neuroinduction is waiting.

But whether you join my programs or chart your own path forward, remember this:

You didn't read this book to stay the same. You read it to become.

The bots are ready. The tools are waiting. The momentum is building.

Your next chapter isn't just waiting—it's already begun.

Now go make it fun—and unforgettable.

— **Dr. V**

Ready to start your AI integration journey?
Visit **funumental.com/talk-human-to-me**

Explore all Funumental programs: **funumental.com**
For the deepest transformation work: **neuroinduction.com**

Made in the USA
Monee, IL
04 November 2026

67387614R00118